THE ASSESSMENT

-The Arrival of Extraterrestrials-

By Adrian G. Rudnyk

ISBN-13: 978-1-7368860-1-4

PAPERBACK PRINT VERSION

FIRST EDITION

To Marian Rudnyk and Taras Matla

We should not listen like children to their parents.
-Heraclitus (540-480) BCE-

Science without philosophy is reckless, and philosophy without science is blind.
-Adrian G. Rudnyk-

Subjectivity must be starting point of any rational inquiry, because with absolute objectivity rational inquiry ends in infinite regress with the "how do you know" game. Hence, rational inquiry simply implies being objective within the context of reasonable assumptions.
-Adrian G. Rudnyk-

CONTENTS

PREFACE

The questions and issues that this work deals with are inherently philosophical in nature that require proper philosophical methodology. Philosophy is a rational system like science is a rational system, albeit a different one. We, philosophers, spend a great deal of time studying logic as any philosophy student will attest. Good reasoning is not just about simply studying deductive and inductive systems. For example, deductive reasoning only preserves truth value, i.e., if the assumptions are wrong, then the conclusion(s) will be as well. Indeed, how we reason is necessarily reflective of our values, our knowledge, and our way of being, i.e., ethics/politics, epistemology, and metaphysics (i.e., ontology). While this work appeals to science and other areas where appropriate in dealing with some of the issues involved, it is, nevertheless, strictly philosophical. There is an important distinction between doing science and appealing to science. The latter does not necessarily imply the former. While there have been many discussions on the kinds of questions this work investigates, unfortunately, those who have commented on such issues have done so in a cursory or ad hoc manner. Both philosophy and science are both rational systems, and each system has its strengths and weaknesses. Science without philosophy is reckless, and philosophy without science is blind. The role of philosophy is not necessarily to provide us with definitive answers to life's important questions or challenges, nor is it to replace science or other disciplines. A philosopher's project is either descriptive, prescriptive, or sometimes a combination thereof. Sometimes, the role of philosophy is to prepare a conceptual framework in which the various disciplines within science then can step in and work in tandem with each other in properly dealing with the challenges presented. Hence, both the rational systems of science and philosophy are needed in dealing with subjects like extraterrestrials and other extraordinary phenomena. My hope is that this work provides a constructive backdrop with respect to the subject of UFOs and aliens.

1. INTRODUCTION

The focus of this work deals with two key questions:

1. *What are some key basic reasonable assumptions that can be asserted about an alien civilization that not only has the capacity to travel to Earth with relative ease, but actually arrived at our planet?*

2. *What would the nature of a relationship between humans and aliens be upon their arrival, i.e., would a friendship develop between our two civilizations?*

For the purpose of this work, there are four conditions that are to be assumed when dealing with the above questions.

The first condition is that the reader should suspend any predisposition as to whether aliens have, or have not, visited Earth. Ignore any connected data and/or experiences. Pretend that you have never before thought about extraterrestrials or UFOs until now. The reason is that people who assert aliens are here have been exposed to conflicting data as to the behavior of aliens, their motives, and their origin. Some of the cases and data can be, and have been, misrepresented and also subconsciously, and even consciously confabulated. What compounds the challenge of interpreting the data, even credible data, is the issue of underdetermination, i.e., different theories can be supported by the same set of facts or the same empirical data. For those who do not endorse the view that there is an alien presence on Earth, this, of course, is a non-issue.

A second condition is a proper understanding of the term *alien*. For the purpose of this work, I take the term *alien* to mean a conscious sentient non-human biological entity from another planet outside of our solar system. By extension, an *alien civilization* would consist of a conscious sentient biologically based species from another planet outside of our solar system.

A third condition is to think of aliens as animals. This seems like an obvious or redundant point to make, given that I have already just defined an alien as biological conscious entity. Nevertheless, it is important to stress this point, because, for example, many humans seem to make a distinction between humans and animals, simply because of our place on the food chain. The point is we are animals and so are the aliens. This consideration is especially important, because as long we remember that the aliens are nothing more than animals, like we are animals, who have worked their way up the evolutionary ladder, then there are reasonable things one can infer about a species, even an alien one, despite what their particular culture might be.

The final condition is that it is an alien civilization that is the focus of this work and not aliens as individuals. The importance of this distinction will become apparent throughout this work. Additionally, please note that given the first condition as elaborated above, this work utilizes hypothetical or future tense language, e.g., 'if the aliens *were* to arrive,' or 'what *should*, or what *would*, we expect from an alien civilization, given X?'

Accordingly, this work is both for the believer and nonbeliever, as it were. With the suspension of a belief that an extraterrestrial presence is already here, a fresh approach is possible in exploring two key fundamental aforementioned questions, which are the focus of this work. While these two questions seem like two wholly separate questions, they are not. They are interrelated in the sense that the second question cannot be properly explored until the first question is answered. Simply put, this work is a philosophical conceptual analysis of these questions. Chapter 2 tends to question one presented above. There are seven propositions that I present and explore in that chapter as reasonable assumptions that can be asserted about an alien civilization that has the capacity to travel to our planet with relative ease. Chapter 3 deals with question two presented above in the manner of applying the seven propositions to the properties of friendship, which will be presented and elaborated within that chapter. As a result, more propositions within this context shall arise. In the end, there will be two sets of propositions: the seven propositions from Chapter 2 and a set of other emergent propositions from Chapter 3. The former are the set of propositions with respect to the first question, and the latter are the set of emergent propositions with respect to the second question.[1] Chapter 4 tends to some possible objections or con-

[1] Please note that there are also arise various corollary propositions to some of the propositions I present. These corollary propositions I refer to as supporting propositions.

cerns one might have with respect to some of the things asserted in this work. In Chapter 5, final conclusions along with some final thoughts are presented.

2. SEVEN PROPOSITIONS AS REASONABLE ASSUMPTIONS ABOUT ALIENS

What are some key basic reasonable assumptions that can be asserted about an alien civilization that not only has the capacity to travel to Earth with relative ease, but actually arrived at our planet?

I assert that there can be seven reasonable assumptions that can be made with respect to answering this question. Accordingly, in this chapter, I shall present and elaborate what I term as the Seven Propositions:[2]

(P1) Aliens would be more intelligent than us.
(P2) Alien technology would be superior to our technology and necessarily different.
(P3) Aliens would most likely be transgenic beings.
(P4) Aliens would be able to communicate with us clearly.
(P5) The aliens' own interests would be their primary concern.
(P6) Aliens would owe us no allegiance.
(P7) Aliens would have good reason to avoid formal public contact with humans.

(P1) Aliens would be more intelligent than us.

Elaboration

While this seems like an obvious assertion, the key question is what is intelligence? Another question is how would we be able to recognize that the aliens would be more intelligent than us? Are certain things we associate with intelligence applicable to aliens? Given the nature of this work, I think (P1) is especially important and therefore, (P1) will be explored more in depth than the other six propositions.

There are many notions of what constitutes intelligence and many aspects concerning with how we perceive, learn, think, and remember. Moreover, I think it is important and incumbent upon

[2] The SP from here on, and referencing the SP implicitly also includes any of the supporting (corollary) propositions within this chapter, which will be presented in due course throughout this work.

me to highlight some of these notions and theories that various experts have purported about intelligence so as to make the point that the topic of intelligence in *cognitive science* [3] is not wholly conclusive and by extension, presents us challenges when assessing an alien intelligence. While there are many approaches to the nature of intelligence and sub-topics with respect to this subject, I shall first, present just a few, yet, important and popular theories of intelligence among scholars. Second, I shall then tend to other issues concerning the nature of intelligence, as well as how they apply to an alien civilization. Accordingly, I shall present some alternative notions of assessing intelligence.

---

[3] *Cognitive Science* is a multi-disciplinary field that studies how we perceive, learn, think, and remember. Some of the disciplines involved are psychology, neuroscience, philosophy, linguistics, computer science, and anthropology.

Some Important and Popular Theories of Intelligence among Psychologists

1. Howard Gardner: Eight Intelligences

Psychologist Howard Gardner asserts that "intelligence comprises multiple independent constructs, not just a single, unitary construct" (Sternberg 507). Moreover, there are eight abilities and each ability is a separate intelligence, and therefore, "not just a part of a single whole" (Sternberg 507-08). Gardner's *Eight Intelligences* are thusly (Sternberg 508):

Type of Intelligence	Tasks Reflecting This Type of Intelligence
Linguistic intelligence	Used in reading a book; writing a paper, a novel, or a poem; and understanding spoken words
Logical-mathematical intelligence	Used in solving math problems, in balancing a checkbook, in solving a mathematical proof, and in logical reasoning
Spatial intelligence	Used in getting from one place to another, in reading a map, and in packing suitcases in the trunk of a car so that they all fit into a compact space
Musical intelligence	Used in singing a song, composing a sonata, playing a trumpet, or even appreciating the structure of a piece of music
Bodily-kinesthetic intelligence	Used in dancing, playing basketball, running a mile, or throwing a javelin
Interpersonal intelligence	Used in relating to other people, such as when we try to understand another person's behavior, motives, or emotions
Intrapersonal intelligence	Used in understanding ourselves—the basis for understanding who we are, what makes us tick, and how we can change ourselves, given our existing constraints on our abilities and our interests
Naturalistic intelligence	Used in understanding patterns in nature

2. Louise Thurstone: Primary Mental Abilities

Another theory of intelligence is by Louise Thurstone. Accordingly, "the core of intelligence resides not in one single factor but in seven such factors" (Sternberg 492). His theory of intelligence is referred to as the *Primary Mental Abilities* (Sternberg 494):

Type of Intelligence	Tasks Reflecting This Type of Intelligence
Verbal comprehension	measured by vocabulary tests
Verbal fluency	measured by time-limited tests requiring the test-taker to think of as many words as possible that begin with a given letter
Inductive reasoning	measured by tests such as analogies and number-series completion tasks
Spatial visualization	measured by tests requiring mental rotation of pictures of objects
Number	measured by computation and simple mathematical problem-solving tests
Memory	measured by picture and word-recall tests
Perceptual speed	measured by tests that require the test-taker to recognize small differences in pictures or to cross out the *a's* in strings of varied letters

3. Robert Sternberg: Triarchic Theory

Yet, another theory of intelligence is by cognitive psychologist Robert Sternberg. His theory of intelligence referred to as the *Triarchic Theory* and all these 'thinkings' work in concert, according to Sternberg (509-10):

Type of Intelligence	Tasks Reflecting This Type of Intelligence
Analytic thinking	solve problems by using strategies that manipulate the elements of a problem or the relationships among the elements(e.g., comparing and analyzing)
Creative thinking	solve new kinds of problems that require us to think about the problem and its elements in a new way (e.g., inventing, designing)
Practical thinking	solve problems that apply what we know to everyday contexts (e.g., applying and using)

4. The Cattell-Horn Theory of Fluid-Crystallized Intelligence

Initially, this theory was developed by Raymond Cattell and developed further by his student John Horn. Cattell identified that there are two types of intelligences: *fluid intelligence* (symbolized as *Gf*) and crystallized intelligence (symbolized as *Gc*). *Gf* is the capacity of reasoning in solving problems in novel situations, independent of acquired knowledge. On the other hand, *Gc* is the capacity of reasoning in solving problems, which is predominantly dependent on acquired knowledge. Additionally, the *Cattell-Horn theory* asserts that intelligence incorporates "100 abilities working together in various ways in different people" (Cattell) within the context of both *Gf* and *Gc* (Sternberg 494-95; Cattell).

5. Other Theories and Approaches to Intelligence

There are many other theories such as Guilford's *Structure of Intellect (SOI)*, Spearman's *"g" Factor*, and *IQ (Intelligence Quotient) Test* that is designed to test intelligence of an individual, and so on.

Some Challenges in Assessing Intelligence

I shall briefly address two kinds of intelligences, linguistic/verbal intelligence and musical intelligence, from the above theories as to underscore the general point that our notion of intelligence cannot be universally applied and therefore, limited in its scope.

Let us first consider the *linguistic/verbal fluency* intelligences. What if aliens are telepathic and communicate in a fashion we have yet to conceive? We must be open to such ideas.
It is a scientifically accepted and publically well-known fact that on our planet there are some animals that, by definition, have extra sensory perception. For example, various aquatic animals like sharks have what is called electroreception. They can sense the variations within the electrical field for locating of their prey (Colla). So by definition ESP is proven as fact on this basis alone. Given this as true, then the possibility of alien life that is very different and more intelligent than us is a plausible idea. Accordingly, we must be open to the notion that there are ways of communicating and perceiving that our science has not proven or yet fully comprehended.

If a form of telepathic communication actually does exist, then that would put into question the validity of verbal fluency and linguistic intelligence as a way of gauging an individual's intelligence. For example, telepathic communication might not involve sentences, i.e., telepathic conveyance of feelings, impulses, and images all at once as communication, and words (not sentences), if even necessary, might only be utilized as indicators or supplements to the aforementioned telepathic message. For example, if an alien desired an apple that is sliced in half, then that alien would merely actually state the word "slice," or telepathically state the word "slice," and then, concomitantly telepathically send an image of a sliced apple along with feelings of desire of a sliced apple as a message to another individual. Perhaps, no words would even be necessary. An image along with feelings of desire or aversion about an object would be telepathically conveyed to another individual, informing them of their desire for a sliced apple. Hence, the image would be the subject at hand and the rest would be the predicate indicating what should be associated with the image.

What of *musical intelligence*? There are some reasons that come to mind why musical intelligence would not necessarily be a proper way in gauging an alien's intelligence.
First, it could simply be the case that there are aliens who do not value music (or sound art at all), not because they would necessarily be deaf, or that they would not have the capacity to compose, perform, or appreciate music, but, simply, because they would not perceive any practical reason for music and its associated activities. As counter intuitive as this may seem to us, it must be considered. However, it could be the case that an alien civilization would have a functional notion of sound, even if, that civilization would not subscribe to music as such. Or perhaps, a civilization could have both a functional notion of auditory qualities of sound and silence as such and a functional notion of music. By functional, I do not mean this in the obvious sense of its meaning such as music for ceremonial sounds for various occasions, e.g., music for weddings, bells ringing after a wedding ceremony. By functional notion of music or just sound/silence as such, I mean a deeper and different philosophical notion with respect to how we associate ourselves with music or just the auditory qualities of sound/silence and how it affects our well-being as it pertains to our moral behavior or even our conscious well-being. Perhaps, an alien civilization could have knowledge about auditory qualities that we have yet to understand with respect to the universe and our conscious well-being or consciousness itself. Perhaps, an alien civilization has developed a whole science supporting a notion that one's exposure to certain auditory qualities or certain kinds of music is a means of cultivating and maintaining a balanced mind and/or even one's body as a whole, thus, allowing one to be in harmony with the universe. I shall highlight four examples as precedent that broadly exemplify the notions just described by appealing to the philosophies of Motzu, Pythagoras, Plato, and Hinduism.

Motzu[4] (480-390) BCE was a Chinese philosopher who advocated the abolishment of music. The core of Motzu's philosophy boils down to ten doctrines (Fraser; Loy). [5] While Motzu wrote

[4] Or Mozi. Motzu or Mozi meaning Master Mo (Loy; Mozi 59).

[5] Motzu's teachings are in a book entitled the *Mozi* consisting of 71 chapters. The ten doctrines sometimes referred to as the ten triads consist of chapters 8-37 within the *Mozi* book. Each doctrine consists of three essays and each of

nothing, the ten doctrines are considered the core chapters and best representation of Motzu's teachings, given that they "derive either from the teachings of Mozi himself or from the formative period of the Mohist community" (Loy). It is from these ten doctrines from that I base my views about Motzu's moral philosophy with a special emphasis on the ninth doctrine.

In moral philosophy, Motzu is what philosophers refer to as a utilitarian. Simply, utilitarianism in moral philosophy means that an act is morally right if and only if it maximizes net utility that benefits the majority. Net utility is a good; something that benefits society that should be maximized. More specifically, many philosophers refer to Motzu as a language utilitarian ("Philosophy of Language"). For Motzu, the primary function of language is to make distinctions. This distinction making power in language is what he refers to as *bian*. There are two distinctions that can be made in language, i.e., *shi* and *fei*. *Shi* is the distinction of "is this" or "right", and *fei* is the distinction of "is not this" or "wrong" ("Philosophy of Language"; Fraser; Mozi 65 footnote 12). In Motzu's view, people have a natural disposition for understanding of how to behave correctly due to their natural instinct to maximize benefit and minimize harm as is self-evident once one has the correct understanding of the usage of language ("Philosophy of Language"). In addition, the application of the three standards of the use of language provide additional guidance for correct moral behavior, i.e., basis of asserting a proposition, verifiability of a proposition, and the applicability of a proposition (Mei 410; Mozi 111). In utilitarian fashion, there is also an aspect of impartiality (i.e., universal love; equal concern for everyone) in Motzu's ethics in that language and associated behavior should be in the context of what benefits society first before one's own interest (Loy; Mei 410; Mozi 60, 68-76). Finally, Motzu's philosophy asserts that there are three basic goods or benefits that should be maximized for society: wealth, order, and population (Mozi 60-61). Anything that interferes with achieving these benefits should be discouraged.[6]

---

the three essays correlates with the title or thesis of that doctrine. Accordingly, the ten doctrines are considered the core of Motzu's teachings (Fraser; Loy).

[6] Some philosophers contend that perhaps Motzu's moral philosophy is more akin to divine command theory, i.e., simply, justification of the correctness of a moral act is based on what a divine power would intend it is. While I appreciate this perspective, Motzu's philosophy merely points to Heaven as a way of demonstrating that the Heaven's views coincides with his, and not that Motzu's justifications are based on Heaven's intentions as such. I do concede that he was not the first to explicitly present a systematic formulation of utilitarianism as such in the manner of Jeremy Bentham, however, Motzu's moral philosophy, at its core, is a form of utilitarianism. In my view, Motzu's moral philosophy should be considered the first documented evidence of ethics that is reflective of utilitarianism. At any rate, most philosophers do view Motzu as a consequentialist more than a divine command theorist.

It is the ninth doctrine that is of interest here, which not only perfectly exemplifies Motzu's form of utilitarianism but also his notions on music in relation to society. Motzu's ninth doctrine is entitled, "A Condemnation of Musical Performances" (Mozi 105-110). Simply, this doctrine advocates the banning of all musical performances. However, there are three supporting propositions one can extrapolate from this ninth doctrine:

- Production of musical instruments is a danger to society.
- Attending musical performances is a danger to society.
- Performing music is a danger to society.

In short, Motzu's philosophy asserts that these implicit propositions are true, because they would interfere with attaining the three basic goods or benefits for an ideal society. In other words, one should not participate in the production of musical instruments, one should not attend musical performances, and one should not participate as a musical performer, because valuable time and resources would be wasted in such pursuits, and as a result, would not maximize that which would benefit society, i.e., wealth, order, and population (Mozi 60-61, 105-110).

It is not an inconceivable idea to imagine that an alien civilization might have a view of music and its place in society in a similar fashion as Motzu's philosophy. Indeed, this is just one such case that have I presented. There are other such cases. For example, Buddhist monks, for different reasons than Motzu's, are not allowed to listen to music or to enjoy any sort of entertainment, given the seventh of the ten precepts in Buddhism, i.e., "I undertake the precept to refrain from dancing, singing, music, going to see entertainments" (The Buddha).

Ancient Greek philosopher Pythagoras (570-497) BCE had a functional notion of both music and of auditory qualities as such, i.e., of both sound and silence. Pythagoras wrote nothing due to the fact that he prescribed to an oral and secretive tradition of teaching his philosophy in the school /society that he formed in Crotona in southern Italy (Guthrie 37). What we do know comes from a movement started by Pythagoras' followers after his death by his original students referred to as Pythagoreanism,[7] which lasted up until the mid to late fourth century BCE (Huffman, "Pythagoreanism"; Guthrie 37). It is during the time of this movement that literature about Pythagoras' philosophy started to emerge (Kolak et al. 15; Guthrie 37). During the existence of Pythagoreanism, a division among the followers occurred due to Pythagoras' passing, and the persecution they endured in which the survivors ended up being scattered to different parts of the Greek world (Kolak et al. 15; Guthrie 37). As a result, two main camps emerged within the Pythagoreans: the *mathematikoi (or mathematici)*, and the *akousmatikoi (or acusmatici)* (Iamblichus 42; Kolak et al. 15; Guthrie 37). The former group concentrated on Pythagoras' teachings on "mathematics, music, and astronomy," and the latter group concentrated on the "soul and the right way to live" (Kolak et al. 15-6). Each group attributed their view as the correct way in understanding Pythagoras (Huffman, "Pythagoreanism"). The division of his students should not be a surprise, given "the universal genius of Pythagoras," because for Pythagoras, science and religion should be considered as two aspects of the same reality (Guthrie 37). Consequently, it is the synthesis of both the *mathematikoi* and the *akousmatikoi* that present us a reasonable depiction of Pythagoras' philosophy.[8]

Pythagoras is, of course, well known for the so called *Pythagorean Theorem*. However, he also formally elaborated on the notion that there is an objective mathematical ratio between intervals of sounds that he considered consonant,[9] i.e., octave as 2:1, a fifth as 3:2, and fourth as 4:3 (Guthrie 38; Iamblichus 62). The exact details of the process that lead him to assert conclusions about ratios between certain sounds are illusive, despite some interesting anecdotes. The most reasonable scenario that many scholars assert is that Pythagoras utilized a monochord[10] or something akin to that, which allowed him to experiment with the length of a musical string with respect to certain sounds and their mathematical relations to each other. It is suggested that Pythagoras experimented with various other musical instruments as well (Iamblichus 69, 229). Accordingly, utilizing a string on a monochord, half the length of the string would be an octave, 3/4 length of the string would be a fourth, and 2/3 length of the string would be a fifth (Russell 21).

---

[7] Pythagoreanism and Pythagorean also refers to the philosophy of Pythagoras during the time of his life.

[8] In the first century BCE there was a resurgence of Pythagoras' philosophy that flourished into the second century CE that is referred to as Neo-Pythagoreanism. (Guthrie 39; Reese 623). Neo-Pythagoreanism was an amalgam of early Pythagorean, Platonic, Peripatetic, and Stoic philosophies (Guthrie 39).

[9] Tones when heard together sounded pleasant.

[10] "An instrument of ancient origin for measuring and demonstrating the mathematical relations of musical tones and that consists of a single string stretched over a sound box and a movable bridge set on a graduated scale" (https://www.merriam-webster.com/dictionary/monochord).

From the interval of a fifth he constructed the diatonic, enharmonic, and chromatic scales (Iamblichus 64-65). At any rate, scholars agree that it is reasonable to credit Pythagoras of being the first individual to formalize an objective mathematical relationship between tones, which in turn allowed him to formulate the aforementioned scales. Moreover, Pythagoras concluded that since the relationship of tones and musical scales could be understood objectively using numbers, then so the universe could be understood in terms of numbers (Aristotle, "Metaphysics" 986ᵃ31-33; Guthrie 38). According to Aristotle, the Pythagoreans thought that "the whole of heaven to be a musical scale and a number" ("Metaphysics" 986ᵃ2), which alludes to what is popularly known as the *music* or *harmony of the spheres* (Guthrie 39). According to Pythagoras, the stars (i.e., the Sun, the Moon, and other planets) by virtue of their orbital movement produced beautiful harmonious sounds, which the human ear cannot hear, "because it has been with us from birth, and sound is perceptible only by contrast with silence" (39).

It is through this objective mathematical relationship among certain consonant tones that lead him to believe that listening to musical pieces that emphasized these ratios of tones would be beneficial for one's health and character (Iamblichus 88). Pythagoras felt music without words was a good remedy against "passions of the soul" like despondency, lamentation, rage, and anger (59, 61). That is to say, music was utilized for molding one's character and maintaining the health of the soul. While, Pythagoras utilized various methods and strategies for himself and his followers for shaping one's character, such as a strict diet of certain kinds of foods, it is the auditory qualities of sound and silence that are of interest here. For Pythagoras being able to remain silent was an important way of practicing self-control (Huffman, "Pythagoras"). Moreover, new initiates to Pythagoreanism had to observe five years of silence (Iamblichus 38). With respect to sound, he utilized musical pieces as a way of shaping one's character and being healthy, as suggested above. For Pythagoras, music was to be used in shaping one's soul, so that one would be resonant or "be harmonious with the heavens" (Collinson 9). It is the harmonious sounds that are the underlying reality of the cosmos, and numbers allow us to objectively decipher and understand the ratios of tones, which in turn would allow us to use these tones so that we could be in harmony with the Universe.

Plato (427-327) BCE also focused on music and he had strong notions about the role of music with respect his ethical system. Being a virtue ethicist, he too, believed that the shaping of one's character was key in leading a moral life, i.e., a happy life and a just life. Plato did not define happiness in terms of one's personal freedom, but more in terms of a person's role or function in society that is in line with that person's natural aptitude, e.g., being a farmer or craftsman that a society needs (Plato 370a-374d). Plato's philosophy asserts that the mind (or soul)[11] has three parts: reason, emotion, and desire. There are also four cardinal virtues: wisdom, courage, discipline, and justice. A virtue is a disposition that enables a person to function well. A person must cultivate the three parts of his/her mind with the three of the four cardinal virtues, in which the fourth overarching virtue would emerge. To reason well one should reason with wisdom. Accordingly, reason should regulate both emotions and desires. When reason regulates emotions, then one can be courageous within one's emotions. It is courage that allows one to "carry things out" (442b9). By allowing reason to regulate emotions and desires, then discipline is achieved. When this harmony of the mind is achieved within an individual, then justice (the overarching virtue) would then emerge and a person would become a just person. Hence, to promote a just society one should properly cultivate one's mind in such a way that would enable one to be internally balanced, which in turn would allow one to enact justice in the world (442a-444a).

According to Plato, the kinds of modes (scales) that musical works are based on that an individual listens to shapes that individual's character by virtue of shaping the mind. This of course suggests that certain music should altogether be avoided. For example, while music based on the Dorian and Phrygian modes would be acceptable to listen to, music based on the Ionian and Lydian modes should be avoided, because they can lead one to drunkenness and lamentation, respectively. That is to say, the Ionian mode is used for drinking songs, and the Lydian mode is used for sorrowful songs, according to Plato. Of course, this suggests that being drunk or being in a sorrowful state of being imbalances the mind. This also strongly suggests a functional approach to music in the deep sense in that listening to appropriate music aides with the cultivation of one's character; the kind of music one listens to affects the ontological status of one's soul, which in turn affects one's behavior. So, while music is permissible for listening, the kind of modes that a musical work would be based on would greatly affect the balance or harmony

[11] Mind and soul are terms that are synonymous in Plato's philosophy.

among the three aspects of the mind as they relate to the cardinal virtues respectively, thereby, affecting the overall character of an individual, which in turn collectively influences the society that the individual lives in (Plato 398c11-399c5).

With respect to written sacred texts known as the Brahamanas and the four main Vedas, Hinduism has its roots going far back as 1,000 BCE to even 1,500 BCE respectively (Ludwig 70).[12] However, it is with the sacred text known as the Upanishads written between 800-200 BCE that consolidated and collectively completed the preceding texts, i.e., the reality of Brahman and atman, the relationship between karma and samsara, and moksha as the goal in life, which are the fundamentals that all Hindus recognize (72). I would also add that the concept of OM not only is crucial, but also is emphasized within the Upanishads. To properly appreciate the concept of OM, one should have a basic concept of monism within Hindu philosophy.

The main thesis of the Upanishads is with its emphasis of defining monism within Hindu philosophy. Hinduism's notion of God is not a monotheistic one but a monistic one; the universe or "all reality is one unified divine reality", i.e., Brahman[13] (Ludwig 510). The various gods within Hinduism are mere aspects of Brahman. This is why Hinduism is not a polytheistic religion but a monistic one. This unified divine reality is made of atman and maya. The atman is simply the conscious self or a collective consciousness selves of the non-material world, if you will, from which we all emerge (91). Our collective souls are the emanation of Brahman. Yet, ultimately, there really is only one consciousness, i.e., Brahman. Maya is simply the material phenomenal and psychological world we experience that is an illusion (Ludwig 93; *The Upanishads* 13). It is not that the material phenomenal world does not exist, but only that it is "something that passes away and is not eternal reality" (*The Upanishads* 13). Understanding and meditating on the concept of OM aides one in being in touch with the ultimate unified reality that is Brahman.

While Hinduism allows for the listening of music, it is with a sound and silence as such that has deep philosophical meaning in Hinduism with respect to Brahman and the individual. In particular, it is with the concept of OM, or AUM[14] in Sanskrit, that it is defined and explored within the

[12] There are four main Vedas: Rig Veda, Sama Veda, Yajur Veda, and Atharva Veda. The four Vedas are also referred to as the Samhitas (Ludwig 66).
[13] In the Upanishads the words *Brahman* and *God* are used interchangeably.

Upanishads. OM is a mantra[15] that is a representation of the cosmic auditory qualities of sound and silence of the divine universe in which the actual sound cannot be heard by the human ear but can only be felt. Meditating on OM is not just a silent activity, but also uttering "AUM" in a long monotonic quasi singing voice repeatedly is a way of harmonizing with the ultimate reality (i.e., Brahman), thereby, coming to true knowledge of the divine reality by feeling of experience that is ineffable, yet, nevertheless, understood.

With respect to AUM, the first letter "A" corresponds to the waking state when "consciousness is turned outward" (Ludwig 95; *The Upanishads* 83). [16] The second letter "U" corresponds to the dreaming state when consciousness is "turned inward but still not unified" (Ludwig 95; *The Upanishads 84*). The "M" is the third letter representing deep-sleep state of consciousness that represents the "blissful, unified, massive consciousness," but not the final goal (Ludwig 95; *The Upanishads 84*). The final state of consciousness, AUM represents the silence that arrives after the "M." This fourth state of consciousness means that one experiences the real atman, because it is the silence that precedes and follows the uttering of AUM (Ludwig 95). It is in this state that one experiences true atman as Brahman (Ludwig 95; *The Upanishads* 83-84). As the Maitri Upanishad states, verse 6.23: "God is sound and silence. His name is OM. Attain therefore contemplation – contemplation in silence with him" (*The Upanishads* 102). Moreover, it is in this sense that sound and silence are to be thought of as one, because sound and silence represent one unified reality that is Brahman. It is in this sense that the mantra OM is the cosmic auditory quality of the universe, and that the true knowledge of all reality is ineffable, yet, understood, by virtue of meditating on OM, and in turn aides one in becoming a truly moral person.

Hence, this functional notion of sound (and silence), is an important process in achieving an enlightened state of consciousness in Hinduism. Interestingly and importantly, it should be mentioned that the concept and mantra of AUM has been adopted within the traditions of Buddhism and Jainism, although, obviously, their notions of the self and ultimate reality differ.

---

[14] I like the spelling of AUM, given the reasoning of the sound; however, it must be noted that phonetically AUM is pronounced OM with the O being long sound, because in Sanskrit A and U together makes the long sound O.
[15] Mantra is a sound, word, or phrase used as a way of evoking a spiritual meditative mental state.
[16] More specifically, within the Mandukya Upanishad.

It is common knowledge that many indigenous peoples throughout our history performed chants, and rhythmical beats, instrumental music, or reiterating of certain tones with the intention of initiating meditation in order to achieve a higher mental state. With respect to understanding our ancient heritage of the utilization of auditory qualities, there is an emerging academic discipline referred to as archaeoacoustics. Archaeoacoustics is sub-discipline within archeology in which ancient sites and artifacts are studied from an acoustical perspective. The Hal Saflieni Hypogeum in Malta, an ancient site estimated to be in existence since the fourth century BCE, is a perfect example of how the ancients utilized acoustics to effect one's mental state. Laboratory tests have confirmed that the use of certain frequencies that resonate within the Hypogeum in fact do "have a strong effect on human brain activity" (Debertolis et al 59). Even today, there are plenty of auditory recordings one can find on YouTube, which are comprised of vocal tones (without words) or instrumental tones, arranged in certain ways with the idea that listening to these tones should promote a certain more enlightened mental state, e.g., binaural tones or isochronic tones. Moreover, we have people who are musical therapists, i.e., music used as psychological therapy. What is interesting to emphasize as suggested above is that many of the ancient moral philosophers and philosophies suggest the notion that morality and cosmic order are intertwined, and exposing oneself to either music or certain auditory qualities is an important process in becoming balanced and therefore, a moral person.

Given the above as true, then it is not inconceivable that perhaps an alien civilization could perceive and utilize music or just certain auditory qualities from a purely functional position. Furthermore, they might have perfected the kinds of things illustrated above into a science, i.e., the science of music and/or the science of auditory qualities for mental and biological health benefits.

Their knowledge of auditory qualities might not seem complex or sophisticated with their use of counterpoint or harmony as with, for example, Bach's fugues or complex harmonies of Chopin's nocturnes for piano; nevertheless, they could utilize arrangement of auditory qualities that would be able to modify our biology and our whole well-being. By extension, such aliens might view the task of becoming a virtuoso musician or gifted composer as we understand it as a complete waste of time and resources. This would be something that Motzu (and probably Pythagoras and

Plato as well) would most certainly agree with, given their philosophical notions of music with respect to morality.

Consequently, to assume aliens would exceed in musical intelligence might not be appropriate in gauging their intelligence (or even our own intelligence, given the examples just presented). These are just two basic examples of intelligences, i.e., linguist/verbal and musical, which I presented that indicate that the above approaches of assessing intelligence are not necessarily tenable due to the lack of universality, i.e., not necessarily applicable to sentient and very different non-humans.

Cognition Instead of Intelligence

There are a growing number of psychologists who assert that perhaps the word *intelligence* is not an appropriate term in judging mental activity of an individual. For example, Dr. Justin Gregg, a psychologist and Research Associate with the Dolphin Communication Project, states:

> In an ideal world, we stop using the word "intelligence" and we just talk about cognition. Cognition is just the things that brains do—information processing leading to behavior. And so we can talk about all the things that, for example, a squid can do in its own little squiddy world, all its kinds of behaviors and the way it thinks about and perceives the world, and then you just get a big long list of squid stuff happening. (Gregg)

Even the actual process of how cognition occurs is up for debate. For example, philosophers Andy Clark and David Chalmers, who specialize in the philosophy of mind, wrote a fascinating essay entitled, "The Extended Mind." They present persuasive arguments for a view known as *active externalism*, i.e., that a cognition is in not exclusively in the head, but that "the environment is driving cognitive processes" (Clark et al. 643). So, while consciousness might be inside of us, cognitive processes do not necessarily reside exclusively within the body.

A basic example of this are people who rely on notebooks for addresses, phone numbers, or directions to certain places as part of the cognitive process of memory and belief (647-49). Hence, part of the cognitive process of retrieving a specific memory may involve a notebook, which is outside one's head. It could be an evolutionary advantage that all of our cognitive processes do not have to necessarily occur only within our respective heads but in the world as well (646). It would be quite a task if we relied only on our brain for all of our memories and various other cognitive abilities; hence, it is much easier to utilize the outside world for this. It is not the scope of this work to consider all of the arguments for active externalism that the aforementioned essay presents. For a detailed arguments supporting such notions, it is best to consult the essay suggested. The point is that certain notions we take for granted are not necessarily true but are more complex and different than we might first envision.

My View of Intelligence and Cognition

It seems that there is no overall consensus among scholars with what really constitutes intelligence. Moreover, even the term *intelligence* is put into question by some with a preference for the term *cognition*, instead. I, for one, do not have a problem with utilizing both terms. Without getting lost in the weeds, as they say, I shall present a general overall basic view of my notions of intelligence and cognition. While, perhaps, there can be an overlap as to the meanings of said terms, I do, however, think there is a general distinction between them.

Generally speaking, I take cognition to mean the processing of a specific ability. For example, calculating a math equation is merely a cognitive ability, or having a good memory, or being good in spelling, etc., are all just cognitive abilities. On the other hand, I take *intelligence* to mean that aspect of our mind, which guides our various cognitive abilities with its will and imagination. In addition, intelligence is also about understanding the meaning of why we do what we do, i.e., having a sense of things. It is one thing to calculate $E= mc^2$, but to understand its significance, then applying it in meaningful and imaginative ways requires intelligence. After all, it is just a straight forward multiplication equation at its core. Yet, to come up with this equation and to fully appreciate its meaning, required true intelligence from Einstein. Einstein created the equation and gave meaning to it, i.e., energy and mass are interchangeable.

In sum, a cognitive process is merely processing of data/information, while intelligence is the union of the application and a sense of understanding of cognitive processes in meaningful and imaginative ways. It is not necessarily how good a species' cognitions are as much as it is how a species utilizes those cognitions in imaginative and meaningful ways, which exemplifies intelligence.

With respect to the various above theories of intelligence, some might say that the theories I presented are not appropriate, given that they are not tailored in gauging the intelligence of a civilization, but, instead, these theories gauge the intelligence of an individual. Therefore, a whole different set of criteria should be considered for assessing intelligence, given it is the nature of an alien civilization that is the focus of this work. This point is well taken. However, I assert that it is useful to consider the above theories for two main reasons. First, because it is an important reminder for us that there can be different aspects to what intelligence is, and that the topic of intelligence is not a settled matter among scholars. Indeed, scholars are not even in agreement with what terminology should be used, .i.e., intelligence or cognition. Second, certain aspects of intelligence or cognitions that are important or prevalent within an alien individual might be, by extension, important within an alien civilization. Nevertheless, again, the initial point of order is well taken. I do agree with the basic notion that, if one were to assess the intelligence of a civilization, then one should also apply a theory of intelligence that focuses on the intelligence of a civilization, given that the focus of this work is from this perspective. This would be especially important, if we were in a situation, which required us to quickly ascertain whether aliens were more intelligent than us in general.

Is there such a theory? There is one popular approach. However, it is a physicist, not a psychologist, who came up with a method that effectively measures the intelligence of an alien civilization. It is referred to as the Kardashev scale.

Kardashev Scale

In 1964, an astrophysicist in the former USSR by the name Nicolai Kardashev came up with a method—referred today as the Kardashev scale—that enables us to assess an alien civilization's technological advancement with respect to the amount of energy it can attain for communication. By extension, this approach effectively also allows us to assess the intelligence of a civilization. Kardashev states that there are three basic types of civilizations: Type-I, Type-II, and Type-III (219). While there have been extensions and modifications to this scale by others, I shall mainly focus on the three types as presented by Kardashev with the exception of the additional Type-0 civilization, which is a useful category for it brings into perspective where we stand with respect to the other civilization types (Kaku).[17]

A Type-I civilization is when civilization's planet can harness light energy that falls upon the planet from its neighboring star and can "harnesses the energy output of an entire planet" (Kaku). It is a civilization that does not need to rely on decomposition of dead animals and dead plants for its energy, i.e., fossil fuel and gas (Kardashev 219; Kaku).

A Type-II civilization is when a civilization can harness all the energy from its closest star directly, e.g., like our Sun. Kardashev asserts that a Dyson Sphere could be built to do this, i.e., a device that would encompass the whole Sun and would be able to harness the energy from the Sun for Earth, for example. This civilization would be at least a 1,000 years ahead of us, technologically speaking, and would be able to harness 10 billion times more energy than a Type-I civilization (Kardashev 219; Kaku).

A Type-III civilization is a civilization that is able to harness seemingly an endless amount of energy, i.e., the energy of its galaxy. This civilization would be at least a 100, 000 years ahead us, technologically speaking, and would be able to harness 10 billion times more energy than a Type-II civilization (Kardashev 219; Kaku).

As a whole we are not even a Type-I civilization by Kardashev's standard. We are, as Dr. Michio Kaku asserts, a Type-0 civilization. A Type-0 civilization relies on decomposition of dead animals and dead plants for its energy, i.e., gas and fossil fuel. According to Kaku, we are about a 100 to 200 years away of achieving a Type-I civilization status (Kaku).

Kardashev's notion of correlating how energy is harnessed for communication with the technological status of an alien civilization is a fascinating one. However, in the end, I do not think his

---

[17] Please note, as tacitly indicated parenthetically, that I, in part, appeal to Dr. Michio Kaku for some rudimentary details of each civilization type for better understanding and elaboration of what Kardashev's scale suggests. With the civilization Type-0, I fully appeal to Kaku, given that this is a civilization type not presented by Kardashev; nevertheless, this civilization type is instrumental in gaining a proper appreciation of where we stand as a civilization with respect to the Kardashev scale.

thesis is necessarily an accurate nor intuitive indicator of a civilization's technological status and by extension their intelligence. Given the scenario presented in the question at the beginning of this chapter, how would Kardashev's scale aide us in quickly assessing the intelligence of aliens that would arrive at our planet? In addition, the aliens could send messages in an unconventional manner in which less energy would be required for communication for vast distances than one would think.

There are some other concerns or observations with respect to some of the details of the Kardashev scale. For instance, why harvest energy directly from the Sun with a Dyson Sphere, when the more practical approach would be to figure out how the Sun actually creates its own energy and then do just that. Perhaps, build a fusion reactor, which could be built anywhere, and, in principle, would be mobile. Another observation with the Kardashev scale is that it does not indicate to the reader what kind socio-political status in the broad sense should be associated with a certain civilization type. Indeed, many people have made and continue to make all sorts of claims about the nature of an alien civilization with reference to Kardashev's civilization types. But the fact is Kardashev's paper makes no such reference to what kind of society should be associated with a certain civilization type. I realize that the arguments within this paragraph are not a direct counter to Kardashev's overall thesis as such, i.e., how a civilization harnesses energy for communication determines the scientific and technological advancement of an alien civilization. His thesis is a fascinating one, and one that I still ponder. There is something to it. In addition, the Kardashev scale has inspired important conversations, given the nature of its thesis. In this sense the Kardashev scale is important, useful, and should not be dismissed. Is there another theory that allow one to assess the intelligence of a civilization perhaps even in a quicker and easier manner? The short answer is yes. I formally present a new theory for this work that I have conceived as an alternative to the Kardashev scale that I aptly named the *Rudnyk scale*.

Rudnyk Scale

The Rudnyk scale is another method of assessing a civilization's overall intelligence. This new scale does not measure the intelligence of a civilization by measuring how a civilization harnesses energy for communication, as with the Kardashev scale, but by simply assessing a civilization's transportation system. Moreover, the Rudnyk scale is not focused with the energy requirement for travel, but with a civilization's distance and time of travel from one place to another when observed or deduced through instrumentation. The distance and time required for a civilization to travel upon their own planet and beyond would be a reasonable indicator as to a civilization's overall scientific and technological status, and by extension, this would be indicative of their collective intelligence. In simple terms, the Rudnyk scale lets one ascertain the intelligence of a civilization and that civilization's overall probable nature (i.e., its probable socio-political status) by simply observing how that civilization moves around. While the nomenclature of the Rudnyk scale is obviously inspired by Kardashev, the details are quite different.

A Type-0 civilization is a sub-planetary civilization. It is a civilization in which traveling is accomplished around that civilization's own planet with great difficulty, i.e., slow and cumbersome. Qualitatively speaking, the kind of traveling that would be present in a Type-0 civilization would be akin to what was on Earth before flight technology became the norm, i.e., flight technology would not be present at this stage on any significant or practical level. If one were to witness a civilization that travels upon its own planet surface, as just described, then it would be reasonable to assume certain kinds of things about that civilization.

First, this kind of civilization would have the kind of scientific and technological knowledge, qualitatively speaking, as our ancestors did during this stage of development. Using our history as an indicator, one can assume this type of civilization would lack proper knowledge of electricity, and that the application of electricity would be non-existent in any of their modes of transportation, and by extension, the application of electricity would not be present in other areas of their applied sciences that would also involve communication and medicine, for example. Second, it is reasonable to assume that superstition would be widespread in understanding the world in a Type-0 civilization. The scientific method would be non-existent or at a very basic level depending on the range of development of a civilization within the Type-0 stage. Third, the planet of a Type-0 civilization would most likely not be united. Their socio-political status would be erratic and most likely be violent locally and globally.

Consequently, if a civilization travels upon its own planet with great difficulty as elaborated above, then one should assume the kinds of things just described about that civilization.

A Type-I civilization is a planetary civilization. It is a civilization in which traveling is accomplished around that civilization's own planet with relative ease and normalcy, i.e., flight technology would be the norm. We have experienced this stage for a while now. A significant amount of humans can and do travel anywhere around the planet with relative ease and even luxury. In a matter of hours, a human can be anywhere on the planet. Some people even travel to Europe and back within 24 hours, and the Space Shuttle circled the Earth in about 90 minutes countless times. If one were to witness a civilization that travels upon its own planet as described above, then it would be reasonable to assume certain kinds of things about that civilization.

First, this kind of civilization would have the kind of scientific and technological knowledge, qualitatively speaking, as we do now during this stage of development. Using our history as an indicator, one can assume this type of civilization would possess proper knowledge of electricity, and that the application of electricity would be existent in most of their modes of transportation, and by extension, the application of electricity would be present in other areas of their applied sciences that would also involve communication and medicine, for example. Second, the scientific method would obviously be much more fully developed at this stage. Third, the planet of a Type-I civilization would still probably not be united. Their socio-political status would most likely still be erratic and most likely still be violent locally and globally. That is to say, the status of a Type-I civilization could very well be like it is on our planet, i.e., not wholly united and 'tribal warfare' could still be present at a significant level.

Consequently, if a civilization travels upon its own planet with relative ease as elaborated above, then one should assume the kinds of things just described about that civilization.

A Type-II civilization is an interplanetary civilization. It is a civilization that can, or does, travel to other bodies of its own solar system with regularity, e.g., its moon and to other planets within its solar system. We are at the beginning of this stage. Our astronauts orbited the Moon in December of 1968, and we had actually landed on the Moon by July of 1969, and we have had many missions to the Moon in a relative short period of time since then. In addition, we have had plenty of unmanned missions to other parts of our solar system since the Moon landings. Furthermore, we are on the path in the next few decades of actually going to Mars. Even though we have not actually landed on other planets within our solar system, we do qualify in being labeled as a Type-II civilization. The reason is that the scientific knowledge and the technologies required for us to travel to other planets within our solar system actually exist. We have come a long way since the Moon landings. The only reason we have not actually traveled to other planets within our solar system is due to politics, and not due to lack of scientific knowledge or lack of technological means. In other words, we have the both the scientific knowledge and technological means to travel to planets within our system right now, if we so desired. Once we have traveled to Mars with relative frequency as we have with our travels to the Moon, for example, then we would be a fully-fledged Type-II civilization. By extension, a Type-II civilization would have a higher degree of knowledge of all the sciences than a Type-I civilization. The core of a Type-II civilization's space craft would still be based on chemical rocketry for the most part, but at the height or end of this stage maybe even fission or fusion based propulsion systems would be utilized. Accordingly, a Type-II civilization would be making more progress with respect to fusion energy, although at a rudimentary level. At a Type-II level, there would be improvement with respect clean and green energy, i.e., more efficient and ubiquitous throughout that civilization's own planet would start to emerge. More progress with computers would be made as well. Quantum computers or something akin to it, with respect to its processing power, would become more and more prevalent during the latter part of this stage.

As for the socio-political climate of a Type-II civilization, that civilization still might not wholly be united. The unity in one form or another would be well on its way as it establishes itself as a full on Type-II civilization. Nevertheless, during the first half stage of a Type-II civilization's development, there could still very well be disunity among the citizens of the planet and some considerable violence. Using our civilization as an indicator, it is reasonable to assume that a Type-

II civilization could still have considerable violence within its society; although, it is during this stage that things would start to change for the better, assuming a civilization would continue to evolve towards a Type-III civilization. This civilization would start to become less and less violent as it slowly evolves towards a pre-Type-III civilization.

Consequently, if a civilization can, or does, only travel to other planets within its own solar system as elaborated above, then one should assume the kinds just described about that civilization.

A Type-III civilization is an interstellar civilization. It is one that can, or does, reach, not only its nearest neighboring solar system, but any solar system within its galaxy. By the time a civilization reaches this level, the technology for interstellar travel would be vastly more advanced and radically different from a Type-II civilization. Any civilization will come to the realization early on, as we have, that real and practical space travel must effectively be exponentially many times faster than the speed of light, which would require a whole different paradigm of how to travel. Notice the word *effectively* in the previous sentence. This is because perhaps the word faster would be the wrong way to think about travel, but the word *different* would be best. That is to say, traveling faster would be understood more in terms how one travels from one point to another with minimal velocity in a brief amount of time as opposed to traveling in extreme velocity that would require a lot of time, as is the case of how we currently travel. For example, a Type-III civilization would be able to bend space-time in some fashion, thus exponentially shortening travel time to vast distances. Energy attained from fusion or antimatter would be easily achieved at this stage, and/or perhaps energy would be harvested in ways we have yet to discover for traveling. A Type-III civilization would have a higher degree of knowledge of all the sciences than a Type-II civilization in the extreme. The technology of a Type-III civilization in all aspects would look like magic to us.

Socio-politically, a Type-III civilization would be fully united. Poverty and war would be non-existent within a Type-III civilization's planet. A society of such a planet would have a vastly different mindset than ours, and a way of living that would be extremely foreign to us. This kind of civilization would have a completely different understanding of nature. Metaphysical questions concerning consciousness and the mind-body problem would be answered by this time. A

Type-III civilization would be a civilization that would be easy to identify, yet, extremely difficult to describe and comprehend.

Consequently, if a civilization can, or does, travel from one solar system to another solar system with relative ease as elaborated above, then one should assume the kinds of things just described about that civilization.

The Rudnyk scale allows for a quick, easy, and commonsensical way of assessing the intelligence of a civilization by simple observation along with common sense deductive inferences without the need of any deep knowledge of math or physics. The Rudnyk scale takes the general approach to travel in which the details from science could be worked out at a later time. The Rudnyk scale would allow our military, for example, to quickly ascertain and recognize that an alien civilization that is able to reach our planet with relative ease would be a Type-III civilization. Also, the Rudnyk scale could be useful to us once we would become a Type-III civilization in which we would be exploring our own galaxy in the fullest sense.

Concluding Thoughts for (P1) and the Rudnyk scale

It is evident, that when discussing the notion of intelligence within this context, it is not just the cognitive abilities that is the issue. Improvements and advancements in traveling necessarily involves the advancement of all the sciences. For example, when traveling, the effects and needs of an entity's body and psychology must be taken into consideration. Traveling also involves knowledge of chemistry and physics, because the craft that would be utilized needs to be built with an understanding of the environment it would be encountering. Technology for communication is important for traveling as was self-evident, especially once we started to travel vast distances with trains and aerial craft, for example. And for scientific knowledge and technology to advance there must be a philosophical disposition that allows for this to occur, i.e., a new worldview; to do things differently one must first think differently. This idea of thinking differently and worldview will be fully elaborated within the context of (P2).

The arrival of extraterrestrials at our planet would not be as simple as meeting beings from another planet that would be just an advanced version of us in some way. It is reasonable to assume that a civilization that would have the technology for interstellar travel between our planet and theirs with relative ease would exhibit physics that would astound us and perplex us. Their craft alone would be indicative of their knowledge of nature (i.e., philosophical and scientific notions) that would astound us, enlighten us, shock us, confuse us, and perhaps even frighten us. To think otherwise is to ignore what true interstellar travel necessarily entails.

Hence, for the purpose of this work I take intelligence to mean that the way an alien civilization is able to move around would be indicative of their scientific, technological, and socio-political status as described above. The rest of this work can be construed as a further investigation and elaboration of the nature of a Type-III civilization as defined by the Rudnyk scale, simply, because by definition, it is the Type-III civilization that actually is investigated throughout this work.

(P2) Alien technology would be superior to our technology and necessarily different.

Elaboration

The key word in this compound proposition is *different.* It is obvious that any alien civilization that can travel from one solar system to another with regularity and with relative ease would exhibit technological superiority over ours. This is not controversial. But what do I really mean by asserting (P2)? In this context, I do not mean superior to mean that the aliens would just be technologically more advanced than us, but, also, be technologically *very different* from us.

There are two senses of the word *different* with respect to technology. The first sense of the term different is with respect to different scientific knowledge and principles related to a technology. A second sense of the word different is with respect to the different kinds of interactions that would be required between an individual and the technology. For example, all things being equal, a gasoline car and a solar car basically require the same sort of *interaction* between the user and the car, i.e., a foot pedal is for acceleration, and another foot pedal for stopping, three basic gears (i.e., drive, park, and reverse), side mirrors, and a rear view mirror. However, the application of *scientific knowledge and principles* from how they obtain energy are quite different.

A gasoline car is primarily powered by energy obtained from the combustion of gasoline. On the other hand, a solar car, while an electric vehicle, is primarily powered by light from the sun.

Let us assume, for the moment, that a solar car that an alien would drive would essentially require the same application of scientific knowledge and principles as our solar cars, however, the interaction in driving the vehicle could be significantly different than ours due to the alien's anatomy and biology, which would be quite different than ours. For example, perhaps the alien's anatomy is such that a car seat is unnecessary. Perhaps the biology of their eyes would allow them to see things in the infrared spectrum. Accordingly, certain things on their car panel could only be viewed properly as long as one would able to see things naturally in infrared mode. Hence, their technology would be reflective of their anatomical and biological needs. Is the alien solar car different than ours? The short answer is no. The application of scientific knowledge and principles would essentially be the same as our automobiles in the sense that both cars would be primarily powered by the light from the sun, because it is the application of scientific knowledge and principles that is the focus here. In this sense, I would not qualify the alien's car as utilizing different technology despite the fact that a different interaction might be required between the driver and the car. I do not mean to imply that a technology cannot be *different* in both senses of the word simultaneously. The general point I am making is, for the purposes of this work, the term *different* (i.e., different technology) should be understood in terms of the application of *scientific knowledge and principles* that a technology is based on, and not the interaction between the user and a certain technology. [18]

Nuclear physicist Stanton Friedman asserts:

> The one underlying crucial fact is that, almost invariably, real technological progress comes from *doing* things differently in an unpredictable fashion. The future, technologi-

[18] I realize that one could say that perhaps the manner of how aliens would acquire energy from their sun could be based on very different kind of scientific knowledge and principles, and therefore, their solar cars could be very different from ours from this perspective. Hence, just because the aliens' cars would be solar based, does not necessarily entail that their technology behind the solar cars would be based on the same scientific knowledge and principles as our solar cars. I agree. However, the example I provided in the thought experiment makes it clear that the hypothetical alien solar car would be based upon the same scientific knowledge and principles as a human solar car so as to better illuminate the two senses of the term *different* with respect to technology.

> cally speaking, is *not* an extrapolation of the past. One has to change how one *does* things. (60)[19]

In his talks and writings, Friedman appeals to the history of science and technology that supports this view. Indeed, history is in fact littered with examples that support the view that technological advancement occurs by doing things differently and not just improvements of the past. For example, when we desired to circle the Earth faster, we built ships that could sail faster, however, in the end, we invented totally different modes of transportation that would allow us to travel exceedingly faster, i.e., airplanes, then jet planes and so on, each in turn utilizing totally different technologies. However, appealing to historical data, while incredibly important, does not give us the *why*. Note, what Friedman states in the above quote, "One has to change how one *does* things."[20] But what precedes the action? What about how one *thinks*? Perhaps a philosophical approach to this issue can provide an answer as to *why* it is that an alien technology would necessarily be different than ours. Accordingly, I present a supporting proposition for (P2):

(P2A) Alien technology would necessarily be different from ours due to their world view.

Elaboration

Thomas Kuhn, a physicist, wrote an influential philosophical work entitled, "The Structure of Scientific Revolutions" in 1962 that focuses on the context of justification, i.e., how scientific theories are justified and how scientific knowledge progresses. Broadly speaking, his thesis asserts that the progress in science is not in small successive increments in which one builds from the other in a logical linear fashion. Scientific theories do not develop in a nice neat linear fashion, because there are psychological, economic, and sociological forces at play, which are buttressed by an existing paradigm, i.e., an established way of understanding of the world. Simply, Kuhn asserts that, in the end, when scientific progress is made, it is in the context of a ridding of an old paradigm with a *new paradigm due to a new way of understanding the world; a revolution in thinking about the world; a new world view* (Kuhn).

[19] Italicized words "doing" and "does" added for emphasis.
[20] Italic added for emphasis.

Accordingly, given the above as true, it is not controversial to assume that a technology a civilization wields is reflective of that civilization's scientific knowledge, e.g., a civilization that has solar panels suggests that the concept of harnessing energy from light is understood by those who built the solar panels. Hence, scientific progress necessarily entails technological progress. By extension, for this to occur an old way of thinking and understanding the world would need to be replaced with a new way of thinking and understanding the world; a new world view. Accordingly, it is reasonable to assume that an alien civilization's technology would not only be different from ours but also be reflective of their world view, which would be quite different than our world view. So, while technological progress comes from doing things differently, it is *the thinking differently* that precedes the *doing things differently*. Once one *thinks* differently, then one can *do* things differently. And in this context to think differently, a new world view and a paradigm shift would be in order so as to be able to conceive and create radically new and different technologies.

It is not a coincidence that radically different technologies not seen before in human kind were emerging once radically different world views were emerging in Europe and the New World. For a time in Europe, before the Enlightenment made the scene, logic and knowledge were politically and religiously centralized and controlled by the Church. However, once the Enlightenment emerged in Europe and the New World, the manner of how knowledge could be gained became less centralized and more secularized, thereby allowing for a new philosophical paradigm that stressed critical thinking buttressed by secularism without any theological constraint, which in turn allowed for the development of the scientific method. Indeed, many things started to change both politically and technologically due to a new way of understanding the world—a new world view.

Scholium

It is not a coincidence that the manner of how electricity was truly understood and harnessed occurred during the Enlightenment and thereafter. While Benjamin Franklin is famous for his experiments with electricity, he, nevertheless, did not discover it. It during his famous experiment in 1752 that he discovered that a small electric spark and lighting were one and the same, when his wrist was struck by a spark from the key on his kite string. Technically, electricity was discovered thousands of years ago around 600 BCE by the ancient Greeks in the form of static electricity when they realized that "rubbing fossilized tree resin, or amber, with animal fur made the resin attract dried grass." The Egyptians, too, were aware of "shocks from some electric fish." *Discovering something is one thing, but to have the mindset in endeavoring to truly appreciate and understand what has been discovered requires a certain world view not just from the discoverer but the respective society and political system as a whole for the discovery to properly take root.* It was an English physicist by the name of William Gilbert, around 1600 CE, who first used the Latin word *electricus* in describing electricity. It was Thomas Browne, author of various physics books, who actually first utilized the English term *electricity* in describing Gilbert's work. Luigi Galvani, an Italian doctor in the 1800s, discovered that when a frog's leg touches two different kinds of metals, the leg would twitch. Accordingly, it was Alessandro Volta who realized the significance of Galvani's findings and "concluded that a kind of electrical potential between two metals caused electrical charge to flow through the frog's leg and make it twitch." In other words, he discovered that there was not only just static electricity, which has been known for thousands of years before, but that there is also dynamic electricity, i.e., moving electrical charge. It is this understanding of moving steady charge of electricity, which ultimately gave rise to many technologies we utilize in our everyday lives, which require electricity, from street lights to microwave ovens, computers, various travelling technologies, communication technologies, various medical devices, weapons, etc. Indeed, Volta was the one who also, not surprisingly and appropriately, invented the first modern battery. Without the knowledge of Volta's conclusions and how to harness electricity I think that we effectively, technologically speaking, would still be in the 19th century (*"Who discovered"*).

Indeed, we are still witnessing the causal effects of this due to the emergence of the Enlightenment with the advent of the radical development of transportation technologies, communication technologies, and medicine. This philosophical shift with how one can gain knowledge allowed for the discovery of quantum mechanics and also the acceptance of re-conceptualization of space and time to space-time, as purposed by Einstein in the 20th century. This gave us a greater insight into our cosmological view of the universe. This radically different way of understanding the universe has and is inspiring scientists in conceptualizing radically different kinds of computers that will make our current computers look like something from the Stone Age in a manner of speaking. Currently, scientists are working on developing computers known as quantum computers, which would incorporate some aspects of quantum mechanics. These quantum computers would operate not on bits in which a bit is in one state *or* another, i.e., a zero *or* a one as todays computers do. Instead, future computers would operate on a notion of a *qubit*. A *qubit* would replace the bit, because, unlike the bit, a qubit would be in two states at the same time, i.e., a zero *and* a one. Quantum computing is still in its infancy, but research is being done for the development of such a computer. Once this is accomplished, this would revolutionize the power of a computer as never seen before. Hence, a more advanced computer is not merely improving its existing system but developing a whole new way of processing and storing data altogether, which are all due to the causal effect of our world view.

Consequently, it is when a civilization is able to have a different world view from its previous one that allows and encourages for the development of new radically different scientific theories and radically different and new technologies. Moreover, it is reasonable to assume that the aliens would have a very different world view than ours, and their technology would necessarily be reflective of this. Indeed, the knowledge of nature that the aliens would bring with them would astound us and perhaps even frighten us.

(P3) Aliens would most likely be transgenic beings.

Elaboration

Transgenic is defined as "being or used to produce an organism or cell of one species into which one or more genes of another species have been incorporated" ("Transgenic"). I would extend the definition to also include the notion that a transgenic being would not just be open to incorporating genes of other species but also enhancing existing ones, and maybe even creating novel ones within this context. Perhaps an advanced alien civilization evolved to such a point that the identity of their civilization would not be solely based on their biology but more by their shared socio-political philosophical principles, which incorporates a transgenic view as I have defined it. It is not to say that an alien civilization might not adhere to a certain biological model, but that the model would be a work in progress as needed. That is, the aliens would be open to modifying their biology, if it were to improve their overall quality of life, or to accomplish some immediate or long term goal that would benefit their civilization. Is there precedence for such a view? I think there is precedence for such a view in the broad sense.

There are humans who espouse to a philosophical view known as transhumanism. Simply, it is a philosophical view that the physical and mental capacities of human beings can evolve beyond their current status by means of science and technology, i.e., either by bio-technological means, artificial technological means, or both. Some humans espouse to a form of transhumanism known as extropianism in which immortality is considered a desired and plausible goal as well (More, "The Extropian Principles"). However, the basic view of transhumanism is that it asserts that we humans can and should increase the boundaries of our own existence by manipulating our genes so that we may achieve a better and flourishing life by means of science and technology. A philosopher, Max More, who is the founder of modern transhumanism in that he first officially formulated a philosophy of transhumanism, defines it thusly:

> Transhumanism is both a reason-based philosophy and a cultural movement that affirms the possibility and desirability of fundamentally improving the human condition by means of science and technology. Transhumanists seek the continuation and acceleration of the evolution of intelligent life beyond its currently human form and human limitations by means of science and technology, guided by life-promoting principles and values. (More, "H+: True Transhumanism")

Max More states that transhumanism is not just about being more than human, but also about taking on fundamental philosophical principles of the Enlightenment, i.e., secularism, reason and science are valued, and where democracy, emotional growth, and goodwill towards its fellow humans are encouraged, and should be developed (More, "Max More – Transhumanism"). If restructuring our DNA in some fashion by adding or modifying our genes would improve us, then it should be implemented. It is a conscious way of facilitating our evolution. The USA is a nation that is not bound by one ethnicity of humans. That is, as long as one aspires to support the Constitution and lives according to its principles, then one can become an American. Of course, our government implicitly assumes that all of its citizens and future citizens would be human. However, if one were to implement a sort of transhumanist philosophy to its logical extreme, then a society could be based upon its values and not solely by the biology of its citizens as such. I think this is the next logical step in our own evolution. Accordingly, would not an advanced alien civilization reach this point in some form or degree within their evolution?

I assert that an advanced alien civilization would incorporate a transhumanist philosophy as the main fabric of its society. Accordingly, instead of the term *transhumanist philosophy*, perhaps a more universal term should be utilized: *transgenic philosophy*.[21] That is to say, evolution would be facilitated through conscious effort by means of science and technology instead of solely relying on the random forces of nature. It is a next logical step for a species to consciously manipulate its own evolution that would be reflected by their values and politics. Once a civilization adopts such a view, then its identity would not be solely based on a certain biological model as such, but, instead, the emphasis of their identity would be more from a socio-political philosophical one; although, their biological model would be reflected through their transgenic values. So in this sense there would still be a biological unity within such a society of some sort; however, given a transgenic philosophical predisposition, biological changes would be inevitable. An alien civilization's philosophy would not necessarily be exactly as Max More's version of transhumanism, but a form of it in some fashion.

(P4) Aliens would be able to communicate with us clearly.

Elaboration

I shall provide evidence in support of (P4) in the context of presenting evidence for the supporting proposition labeled (P4A):

(P4A) It is incumbent upon the more advanced species to figure out how to communicate with the less advanced species.

---

[21] Instead of the term transhumanism, a more universal term would be appropriate: transgenism (i.e., beyond one's existing genes); a philosophical view that captures the essence of a transhumanist philosophy elaborated here. Someone who believes in transgenism would be labeled as a transgenist and believes in a transgenic philosophy.

Elaboration of (P4A)

It is self-evident that we, humans, are the most intelligent animals on this planet. We are the only animal on this planet that have traveled to the Moon and back, and we have sent unmanned craft throughout our solar system and beyond. We practice agriculture, farming, and industry, which allow us to produce food and goods in ways that no other animal on this planet is able to do. Furthermore, the aforementioned indicates that we, humans, actively share intergenerational knowledge and plan for the future in a sophisticated manner, unlike other animals. Another aspect that makes us more intelligent than other animals, and unique on this planet, is that we have accumulated much knowledge (and continue to do so) about many other non-human animals on this planet. We have scientists who specialize in animal communication, animal behavior, and animal psychology and who devote all their time in understanding how and why certain non-human animals behave and communicate the way they do and think the way they do. We also have physicians for the purpose of healing other animals. It is true that other animals might have cognitive abilities that are different than ours, and in some cases superior to ours. For example, Dr. Matsuzawa, of the Primate Research Institute at Kyoto University in Japan, proved that chimpanzees have a superior working memory to that of humans (Matsuzawa). Nevertheless, as a whole, humans are more intelligent than chimps and other animals on this planet as is self-evident. The point is, as a matter of course, we actively study and learn how other animals communicate and behave. Indeed, we are able to communicate with many other animals on a basic level utilizing their type of body language and sounds, for example. We also take the added measure of actually teaching some animals how to communicate with us. For example, let us look at another ape for a moment: the gorilla.

Our researchers know how to approach a group of apes, because they have studied and learned their body language and the kinds of sounds and vocalizations gorillas employ, which would signify danger, acceptance, etc. Gorillas did not teach us how they communicate, we had to observe, study, and learn ourselves, because we are much more intelligent and apt in learning how other animals communicate. We have studied and learned about their society and politics, and we took all of that into consideration when we interact with them. What do the gorillas do in the wild? Do they bother to study us and learn our communication (i.e., body language and vocalizations alike) in the wild? Do gorillas in the wild learn about our social structure so that they may communicate with us better? Do gorillas in the wild share knowledge of us with each other? The short answer is no. Gorillas do not have the natural inclination and nor the ability to learn about us as we learn about them. Consequently, it is us, the more intelligent species—humans in this instance—that had to take the initiative and establish proper communication among us and the gorillas. We have also taken the added measure in ensuring even a deeper level of communication with some gorillas by taking them into captivity and teaching them sign language in a controlled setting. We realized that sign language is a more viable way of bridging the gap between us and the gorillas, given gorillas cannot have a spoken language, in part, due to their limited vocal abilities. We are the ones who deduced this, and we realized that sign language could exponentially increase the vocabulary between us and the gorillas. And, indeed, it has done just that.

The famous gorilla known as Koko is a well-known example of this added measure. We realized that Koko has limited vocalization abilities and cannot really talk, but she can understand spoken language, and with our aide, she learned to use sign language. Koko has learned over 1,000 signs and has an understanding of about 2,000 spoken words in English. How did she learn? She learned from us humans in a controlled environment. If it was not for researcher Dr.Penny Patterson, who spent decades studying Koko and actively teaching her how to communicate so that humans and Koko could converse, there would not be any sort of meaningful and sophisticated way of communicating with the gorilla. Interestingly, after a while, Koko started combining symbols representing objects she did not have signs for, thus, making new combinations of signs on her own. Interactions with Koko have taught us that a gorilla's mind is more complex than we first envisioned. Nevertheless, it was our intelligence and effort that lead to this remarkable interspecies communication between us and Koko, thus, providing a clearer picture of a gorilla's mind. We were able to communicate and teach the gorilla to communicate with us on its own level of comprehension. It was our intelligence and effort that was key for this interspecies communication. Indeed, given our level of intelligence with respect to gorillas, it was incumbent upon us to do so, and we were and are able to communicate with the gorilla clearly and on its own level of comprehension ("Scientific").

It is in a somewhat like manner between an alien civilization and us. We would be to aliens like apes are to us. However, aliens would not simply just travel here and land and then try to communicate with us. They would, instead, for example, be covert and study us. They would tap into our electronic media and learn how we communicate and study our languages, psychology, philosophies, and politics and then, if they so choose to, then they would modify their form of communication that would be intelligible to us. An alien civilization as presented here, would figure this out easily. They would have greater success in figuring out how to communicate with us than we would with them, given their advanced intelligence and capabilities. Indeed, given the analogy of us and apes, it would be incumbent upon the aliens to dumb down to our level than for us go up to their level, because it is easier for a more intelligence species to figure out how to communicate with a species of far lesser intelligence, than vice versa. It is in this sense that it is incumbent upon the greater intelligent civilization to take the first initial steps in communication to the less intelligent species, if communication were to be established. This is the main point of the analogy presented.

In addition, please note that in the beginning of the preceding paragraph, the fifth word is "somewhat." The reason I chose this word is because the analogy of comparing aliens communicating with us would be like us communicating with gorillas is not wholly accurate, even though it is a useful analogy. That is, I think it would be easier for an alien civilization to able to figure out how to communicate with us in the most efficient manner, given that we as a species are already aware and understand the concept of different kinds of communications. We have specialists in linguistics, semiotics, psychologists of language, philosophers of language, scientists of artificial intelligence, etc. We know how to record interactions between us and other animals on various media, so that we would be able to review, study, and learn about those interactions. We actively study the behavior and communication of many animals on our planet as a matter of course. By extension, we would no doubt actively try to understand and communicate with the aliens. We are not passive communicators as apes and other animals are on this planet, because, other things being equal, other animals as a whole, do not actively study and learn our communication or other animal communication, of their own accord, as we do with other non-human animals. For the most part, animals study other animals more in terms of how to avoid or attack each other on a basic primitive level, i.e., for food, escape, or reproduction. Furthermore, we also appreciate and actively explore various non-traditional forms of communications within the arts like paintings, dance, literature, movies, and music. Our science fiction movies entertain the various ways how an alien civilization might communicate with us. It is also important to consider that the aliens, too, would have been well practiced in studying the nature of communication as such as well during the course of their own evolution. In other words, it is reasonable to assume that aliens, too, would have been studying the nature of communication much longer than we have, and that they also probably would have been exposed to various ways of communicating as part of their evolution as we have in ours and, given that they would be a spacefaring civilization, their knowledge of communication would be superior to ours. Indeed, a spacefaring civilization would not venture out to space to visit or study other civilizations, if they did not consider how they would probably have to communicate with other intelligent life forms from other planets, even if only to eavesdrop.

Hence, while, in a certain way we would be to aliens like apes are to us, in the end, it would be easier for an alien civilization to establish communication with us, given the way we (both aliens and humans) actively study communication on a diverse and sophisticated levels as just elaborated as a matter of course.

Consequently, it is in this sense that I assert that not only would aliens be able to communicate with us clearly, but that it would be incumbent upon the aliens to establish communication with us, if communication were to be established, given that the extraterrestrials would be the superior civilization. Aliens, being a more technologically advanced and intelligent species, would be able to adjust their communication in such a way that would enable us to understand them at the most basic and important level clearly.

(P5) The aliens' own interests would be their primary concern.

Elaboration

Any intelligent civilization would consider their own interests first before the interests of others. Indeed, lower life forms consider their own interests first, even from the most basic instinctual level. This seems obvious, self-evident, and not a controversial assumption. However, it is important to take note of that even if there were to be an extraordinary scenario in which aliens were to work with us in the "background" trying to help us in some fashion, it would still be from the consideration that their own interests would come first, even if not overtly stated or indicated in some manner. This is a reasonable assumption, and something one should expect from any intelligent civilization that would have worked its way up the evolutionary ladder.

(P6) Aliens would owe us no allegiance.

Elaboration

Why should aliens give their allegiance to us, given (P1), (P2), and (P5)? The answer is, all things being equal, they would have no reason to. The term *alien civilization* itself is suggestive of a species that is not internally connected nor responsible to us economically, socially, nor politically. Indeed, they would not even be Earthlings. Given what the aforementioned propositions collectively suggest, the motivation of forming an allegiance with us would be greatly diminished. If they would need something from us or our planet, they would just take it, just as we do with lower life forms on our own planet. However, if they were motivated to consider some sort of allegiance, it would be motivated by a need, i.e., their need first, and from purely practical reasons in which having some sort of allegiance with us would make it easier to get what they would need.

Of course, I have heard some assert that for an alien civilization to become so advanced as to become interstellar civilization means that the aliens would have learned the art of cooperation. By extension they conclude that this suggests the aliens would be cooperative with other civilizations, even with primitive humans, and therefore peaceful with high morals. I concede that perhaps an advanced civilization might have learned how to cooperate. However, *just because the aliens would have learned how to cooperate amongst themselves does not necessarily mean they would cooperate with us*, especially given (P1), (P2), and (P5). I label this supporting proposition (italicized portion) as (P6A). (P7), which will is discussed in the next section, bolsters the position of what (P6A) asserts.

Given (P1), (P2), and (P5), and arguments presented, then this should provide one with great pause as to the meaning of (P6) within this context. As a side note, we have to remember that even if an alien civilization wished to ally with us, then who is it they would give their allegiance to, given that our planet is not politically united, i.e., there is no true representative of Earth who can speak for all of us in the truest sense, because we are not truly a united Earth civilization. I invite the reader to take a moment before investigating the next proposition and briefly ponder on the idea (P6) truly entails.

(P7) Aliens would have good reason to avoid formal public contact with humans.

Elaboration

What do I mean by *formal public contact*? Formal public contact would entail an alien civilization formally introducing themselves as to who they are, where they are from, and their general intentions of visiting our planet, and then trying to develop a proper relationship between their world and ours in such a fashion that it would be known to all, i.e., a true genuine friendship that necessarily would entail goodwill. However, given (P1), (P2), (P5), and (P6), the aliens would not willingly divulge who they are, where they are from, and what their intentions would be, nor would they share their intimate knowledge of reality. What motivation would they have to provide us with all of their knowledge? They would have nothing to gain, but plenty to lose, given how dangerous we are as shall be elaborated within this portion of this work. Before aliens would decide to actually make formal public contact with us, it would be reasonable to assume that they would study us for a while, i.e., a survey of our civilization. An alien's survey of us would indicate various things about us, which would deter them from having formal contact with us as defined here.

Aliens would notice that every town and city all over the world has a police force. The police force is not there to guard citizens from aliens, but to guard its citizens from other citizens within the same town, city, or state. We even have the police policing the police, because even the police can be abusive. In totalitarian regimes, like the former USSR, for example, the secret police (i.e., KGB) were used to persecute and eliminate undesirables; anybody at a whim really, so as to instill fear and keep the general population in complete control. Currently, there are still regimes that simply imprison or execute their citizens who simply even question their own government, e.g., North Korea and various Middle Eastern nations. So police are not necessarily there to insure safety of its citizens but to oppress them. Hence, the aliens would realize that humans even mistrust and mistreat each other, given the extreme policing among humans against other humans.

Aliens would notice the type of government systems humans have. That is to say, they would notice that there is no unitary government system, but many competing ones. The aliens would notice that centralized governments tend to be more ruthless to their own citizens and that the governments that are less violent to their citizens have checks and balances within the governmental system as with our nation the USA. All of these checks and balances in our own country of the USA exist simply, because our framers were aware of our corrupt and violent nature. In principle, we should not need a balance of power, but a centralized government. However, given human nature, it would be prudent not to have a centralized authoritarian government as history has demonstrated time and time again. The point is, these are all signs of an unstable civilization. Moreover, the aliens would quickly notice that we wage war as a matter of course. Just consider our recent past in the 20th century.

- The death toll of WWI is about 38,000,000 [22]
- The conservative estimate of the death toll of WWII is about 60,000,000 [23,24]
- The death toll of the Korean War is about 1,200,000 [25]
- The death toll of Vietnam War is about 1,450,000 [26]

That is about 100,650,000 people that were killed by other people in wars in the 20th century.[27] This does not even include the millions who died in the gulags of Stalin and Mao. Let us not forget the manner with how we killed each other, which included the nuclear bomb in some in-

[22] Check Works Cited under "World War I Causalities."
[23] Check Works Cited under "World War II Casualties."
[24] Check Works Cited under "By the Numbers: World-Wide Deaths."
[25] Check Works Cited under "Korean War."
[26] Check Works Cited under "Vietnam War Casualties."
[27] The death toll of the wars presented include both military and civilian, except for the Korean War, which only depicts total of battle deaths.

stances. History clearly shows that humans have turned on each other time and time again. The 21st century has demonstrated, thus far, that our violent ways have not ceased.

Finally, aliens would come to realize that humans not only attack other humans as a matter of course, but a human individual will attack himself, i.e., suicide. In the U.S. alone suicide is listed as the second or fifth—depending on the age group—leading cause of death in 2013:[28]

- Second leading cause of death among humans between the ages of 15-34
- Third leading cause of death among humans between the ages of 10-14
- Fourth leading cause of death among humans between the ages of 35-44
- Fifth leading cause of death among humans between the ages of 45-54

Given these facts, the aliens would have good cause for not having formal contact with humans. The suicide rate and the wars we wage against each other alone demonstrate how highly unstable and dangerous we are. With respect to how violent we are with each other as just presented, the aliens would quickly conclude that: *Given how humans treat each other, there is no reason to assume that humans would treat aliens any differently*. This supporting proposition I label as (P7A).

We as a species, are an extremely unstable and aggressive animal, and it is reasonable to assume that an alien civilization would be all too aware of our violent and unstable nature. Accordingly, after an alien civilization finished surveying our behavior, they would quickly come to realize that formal public contact would be a danger to their own survival and interests.

[28] CDC (Centers for Disease Control and Prevention); Check Works Cited under "10 Leading Causes of Death by Age Group, United States – 2013."

3. FRIENDSHIP WITH ALIENS

Introduction

What would the nature of a relationship between humans and aliens be upon their arrival, i.e., would a friendship develop between our two civilizations?

Some propositions from the previous chapter have already provided some indication of an alien civilization's behavior towards our civilization. Nevertheless, exploring the question in this chapter will illuminate other important aspects of an alien civilization with respect to their attitude and behavior towards us. This portion of the work explores the question above by applying the SP within the context of friendship between two civilizations, i.e., the relationship between an alien civilization and a human one.

This chapter is not a conceptual analysis of friendship as such, nor does this work delve deeply into the properties of friendship as such. The main focus in this portion of this work is to investigate key properties of friendship only as it pertains to the relationship between two civilizations, i.e., an alien civilization and a human one, with the emphasis as to what kinds of behavior we should reasonably expect from the aliens, given certain reasonable assumptions as elaborated within the context of the SP. Nevertheless, I do provide some analogies of friendship between individuals as opposed to civilizations but only so as to underscore certain aspects between aliens and us clearly. Given that the focus in this work is between an alien civilization and a human one, the key properties of friendship that I shall analyze within this context are *communication*, *equality, goodwill*, *loyalty*, *trust*, *honesty*, and *deception*. In addition, these properties are to be understood in commonsensical ways. I do, however, provide clarity and elaborate on certain terms where it is necessary.

Communication

Of course, one reasonable concern with respect to developing a friendship is communication. If an alien civilization and a human civilization cannot communicate with each other, then it would be doubtful that a proper friendship could emerge. How can a friendship develop between aliens and humans, if said parties could not communicate properly? As established within the context of (P4) in the previous chapter, if an alien civilization wanted to communicate with us as a first step for the purpose of establishing a friendship of some sort, then, in principle, they would be able to do so. However, it is doubtful that aliens would formally communicate with humans, given what (P7) alone asserts, but, also, what (P5) and (P6) assert. Consequently, it is in this sense that *the property of communication would not exist between aliens and humans*. I label this proposition (italicized portion) as (EP1).[29]

Equality

Can humans and aliens have a friendship based on equality as true friendship requires? In a relationship based on equality, the parties involved must be equal to each other. This is the basis of a true or genuine friendship that is based on equality be it on an individual to an individual basis, or a civilization to a civilization basis. By true or genuine friendship, I mean that the parties involved would have mutual respect and value each other's rights within their friendship as equals as we would intuitively understand it. By equality, I do not mean absolute equality in which the parties involved are literally equal in every way. For example, equality in this context does not entail a friendship in which cognitive abilities, differing economic backgrounds, or physical strengths, etc. have to be literally equal among the parties involved in a friendship. Equality in a genuine friendship means that the parties involved intrinsically value and respect the other *as* equals despite obvious inequalities in the absolute sense that would exist among the parties involved. In short, equality should be understood in a friendship in a qualitative way and not a quantitative one. This should be understood in a commonsensical and intuitive manner.

However, it is doubtful whether true friendship could exist between humans and aliens, given the SP. In addition, given (P1) and (P2) by definition, aliens would not be our equals in the extreme. Consequently, it is in this sense that *the property of equality would not exist between aliens and humans*. I label this proposition (italicized portion) as (EP2).

Scholium for (EP2)

[29] The "E" is part of the nomenclature because this proposition is an emergent proposition as a result of what was asserted in this subsection and by virtue of what (P1) and (P2) suggest from the previous chapter. From here on all the emergent propositions from this chapter will collectively be referred to as the *emergent propositions* or the EP, and for the same kind of reasoning as just stated in the first sentence of this paragraph.

While equality within a true friendship should be understood as qualitative equality, to what degree should qualitative equality be taken? Can there be a friendship based on qualitative equality, if the difference between the parties involved is too much? Can a friendship based on inequality exist? Indeed, there are various kinds of unequal friendships.[30] For example, a friendship between a parent and a child is an unequal friendship, just like the friendship between the state and its citizens, or a nation with another nation. Interestingly, even in an unequal friendship both parties could truly care for the other. For example, a mother and her child can develop a deep bond with each other. What of an unequal friendship with aliens? Aristotle who wrote an extensive philosophical analysis of the nature of friendship in his work entitled, "Nicomachean Ethics," warns us that if one party in a relationship is too unequal, then there is no possibility of any sort of friendship.[31]

> ...men of no account expect to be friends with the best or wisest men. In such cases it is not possible to define exactly up to what point friends can remain friends; for much can be taken away and friendship remain, but when one party is removed to a great distance, as God is, the possibility of friendship ceases. (1159^{b}1-5)

Aristotle's point is well taken. I would imagine this could be true, because the relationship could be too one sided. Would an alien civilization be like gods to us? It is difficult to say, but not farfetched to imagine. There is no doubt that their technology would be akin to magic to us, and that their knowledge of nature would be vastly different from ours, given (P1) and (P2). For example, I can easily imagine people jumping to a conclusion that aliens, whose craft might appear to pop in and out of existence, are interdimensional, when it could simply be the case that their technology reflects their knowledge of applied physics that we have yet to discover or fully understand. Indeed, people a few hundred years ago would no doubt view our current technology as something supernatural or interdimensional. At any rate, there are some shared properties between equal and unequal friendships. However, (P1), (P2), and (P3) established that an alien civilization would not just be far more advanced than us, but also extremely different from us. Con-

[30] Please note that the utilization of the expression "unequal friendship" is a shorthand for "friendship based on inequality."
[31] All quotes and parenthetical reference referring to Aristotle in this chapter are from his *Nicomachean Ethics*.

sequently, a relationship between an alien civilization and a human one would necessarily be an unequal one.

Goodwill

I take my cue from Aristotle that in true friendship, goodwill should be present in a mutual way such that the parties involved must have feelings of mutual goodwill towards each other, the parties involved should be aware of each other's feelings towards each other, and goodwill should originate within this context (1155^b27-34, 1156^a1-5). I take goodwill to mean that one can have a friendly, kind, and compassionate attitude towards the other party. Could the relationship between aliens and humans involve mutual goodwill as just defined? The short answer is no. As mentioned earlier, the relationship between an alien civilization and a human one would necessarily be an unequal one. Accordingly, the main reasons that there would not be goodwill as described above are due to (P5), (P6), and (P7). There cannot be any feelings of mutual goodwill in which both parties are aware of each other's feelings unless there is meaningful and genuine contact among the parties involved. The aforementioned propositions provide no motivation for aliens to interact with humans in any meaningful way such that goodwill would emerge. (P7) especially discourages any meaningful contact, as explored in the previous chapter that would be required for a true friendship. How would aliens and humans mutually be aware of each other's goodwill, unless there would be mutual meaningful, genuine, and direct contact? Is mutual goodwill required in a friendship based on inequality? Of course, it would be desired property, even within this context. Indeed, there can be mutual goodwill between a child and its parent. However, the lack of feeling mutual goodwill, as just defined, does not preclude the notion that one party could have feelings of goodwill towards the other, even though one party would not be aware of it. More to the point, it does not preclude the notion that an alien civilization could have feelings of goodwill towards us, even though we would not be aware of it. Hence, goodwill, as such, could possibly exist between aliens and humans, even though the parties involved would not be aware of it. But again, this is not true or genuine friendship. For genuine friendship to exist between the parties involved, the parties involved not only must have a feeling of goodwill towards the other, but they also should be aware of their mutual feelings of goodwill towards each other.

Hence, in principle, aliens could possibly have feelings of goodwill towards a human civilization and yet, given (P5), (P6), and (P7), the aliens would still have good reason to avoid us in the formal sense, until such time we have evolved properly for first formal contact as equals. Hence, given (P5), (P6), and (P7), in the end, aliens would not have feelings of goodwill towards humans. Consequently, it is in this sense that *the property of goodwill would not exist between aliens and humans*. I label this proposition (the italicized portion) as (EP3).

Loyalty

Whether friendship is a one-to-one relationship, or civilization-to-civilization, it goes without saying that there should be loyalty among good friends for various reasons, e.g., protection from mutual enemies. This is obvious but important to mention, given it is one of the key properties of friendship, and something that different nations or civilizations would aspire to, if they were to be on friendly terms with each other. Given (P5), (P6), and (P7), aliens would not have any motivation in being loyal to humans. Consequently, it is in this sense that *the property of loyalty would not exist between aliens and humans*. I label this proposition (italicized portion) as (EP4).

Trust

For trust to develop among friends there must be familiarity. Aristotle states this about trust with respect to friendship:

> Further friendship requires time and familiarity; as the proverb says, men cannot know each other till they have 'eaten salt together'; nor can they admit each other to friendship or be friends till each has been found lovable and been trusted by each. ($1156^{b}26$-30)

For genuine friendship to develop, the parties involved should fully be acquainted with each other in the most important ways. This cannot occur unless the parties involved spend time with each other. Given what (P7) alone suggests and especially its corollary (P7A), aliens would be extremely distrustful of humans. Aliens would have no motive in having any meaningful contact with us, which is needed in establishing trust. Additionally, given what (P5) and (P6) assert, this would motivate the aliens even more so in keeping their distance from us with respect to establishing trust. Consequently, it is in this sense that *the property of trust would not exist between aliens and humans*. I label this proposition (italicized portion) as (EP5).

Honesty

When both parties are honest with each other, then a stronger bond can emerge. By honesty, I do not mean that every little detail and secret from both sides should necessarily be divulged amongst the parties involved. By honesty I mean that both parties would be honest in the most basic and important ways: *who they are, what their core basic beliefs are, and their key general desires within the context of equality and goodwill.* Hence, honesty would strengthen the bond among the parties involved, which is needed for true friendship. Can this kind of honesty exist in an unequal friendship between an alien civilization and a human one? The short answer is no.[32]

(P7) and (P7A) alone established that aliens would not have good motivation in spending time with us. It is also true that aliens would be highly distrustful of humans by virtue of (EP5). Given the last two statements as true, then there is no reason to share because there would be nothing to be honest about. Accordingly, aliens would have no motivation in being honest with humans with respect to sharing their scientific and technological knowledge as well as who they are and where they come from. Consequently, it is in this sense that *the property of honesty would not exist between aliens and humans*. I label this proposition (italicized portion) as (EP6).

Deception

Prima facie, it seems strange that deception would be a property of friendship, especially for a genuine friendship. However, do we not spare our friend's feelings by keeping certain things to ourselves until the right time, or perhaps we would never tell our friend certain things, as the

[32] By honest, I mean in the positive sense and not the negative sense. For example, I can be honest in saying I hate X and I wish X would just go away. Those are honest statements, but negative ones. By honesty, I mean the kind of honesty as implicitly understood in commonsensical ways that we would expect from a friendly civilization.

case may be, for a variety of reasons? Sure, we all would, and many of us have done so. All good friends would do so out of compassion motivated by goodwill. In an unequal friendship between a parent and a child, for example, it is a parent's job to ensure that the child feels safe, and to even lie to a child, in order to instill a feeling of safeness to the child when at times conditions might not be safe. [33] In an unequal friendship between the state and its citizens, the state must employ a certain amount of deception as to ensure that the nation remain stable. For example, the state must implicitly assure its citizens that they are safe. The state would not share with its citizens, the methods and technologies utilized for security, so as not inform a potential enemy. In unequal friendships, deception would perhaps be more pronounced. However, the motivations for deception can vary. This concern will be explored shortly.

Given what (P1) and (P2) alone established, the aliens would not be our equals and they would be scientifically and technology more advanced than us. Indeed, an alien civilization might just avoid us altogether, given what (P5), (P6), and (P7) collectively suggest, especially (P7). Let us not forget the established emergent propositions (EP1), (EP2), (EP3), (EP4), (EP5), and (EP6) within this chapter. However, if aliens were motivated in interacting with our planet for some reason, given the aforementioned propositions mentioned within this paragraph, it is reasonable to assume that the aliens' behavior towards humans would necessarily be deceptive in the extreme. Consequently, it is in this sense that *the property of deception would necessarily exist between aliens and humans in the extreme*. I label this proposition (italicized portion) as (EP7).

If the aliens were going to interact with our planet, then their deception would take on one of three basic forms: *the neutral sense*, *the positive sense*, or *the negative sense*. I shall briefly explore each of these three kinds of deceptions with appropriate hypothetical scenarios, which in turn bolsters (EP7). The point of these hypothetical scenarios, which I shall present, is that even at the most rudimentary level, the aliens would necessarily employ deceptive behavior towards us in the extreme, despite what their intentions towards us might be, given that an alien civilization would not be our equals as established by (P1) and (P2) and what especially (P5), (P6), and (P7) collectively suggest. It is in this sense that I assert (EP7) as true. I shall now elaborate on the three basic forms of deceptions mentioned above.

---

[33] Lying is a form of deception.

First Scenario of Deception – Neutral Sense

The core properties of deception in the neutral sense are thusly:

1. There would not be any interaction between aliens and us.
2. The aliens would strive not to help us, nor harm us.
3. The aliens would only be on our planet for some practical reason, e.g., perhaps there is some resource on our planet that is unique in some way.
4. Given what (P5), (P6), and (P7) collectively suggest, the aliens would be motivated in practicing deception to the best of their ability; remain completely mysterious, or hidden from us as to their presence and their intentions.

One possible scenario, within this context, would be that they could simply want some sort of resource that is, in some way, unique, yet, abundant on our planet. One specific reason for their neutral approach of deception could be out of sheer practicality, i.e., they would decide to simply avoid formal contact, remain mysterious to us, and take what they want (or need) from our planet without the trouble of dealing with us, thereby, avoiding the danger we would necessarily pose to them, given what (P7) alone suggests. By avoiding direct formal contact with humans, there is one less thing they would have to deal with while still getting what they need. While we are primitive, we are at the same time extremely clever and dangerous. We can split the atom after all, and as our history clearly has demonstrated and our current state of affairs clearly demonstrates that we still are a warring species as a matter of course. Given the manner with which we treat each other, there is no reason to assume we would treat aliens any differently. It is reasonable to assume that the aliens would arrive to this same conclusion. Hence, one practical approach in dealing with us is deception in the neutral sense. Furthermore, by not presenting a clear and present danger towards us, the human civilization would not be as motivated in a united way in finding out who these aliens are even if their craft were spotted from time to time by radar, pilots, or the general public.

Second Scenario of Deception – Positive Sense

The core properties of deception in the positive sense are thusly:

1. There would be interaction between aliens and us.
2. The aliens would strive to help us instead of harm us.[34]
3. Their behavior would be guided by goodwill; out of caring for our species from an altruistic perspective.
4. Given what (P5), (P6), and (P7) collectively suggest, the aliens would be motivated in practicing deception to the best of their ability; remain completely mysterious, or hidden from us as to their presence and their intentions.

For example, let us imagine one hypothetical scenario in which the aliens took notice that we were stuck in an evolutionary bottleneck due to unnatural circumstances, and felt that the situation should be remedied. The exact nature of what affected our evolution in an unnatural manner is not the point, but only that whatever it would be, the aliens would deem it necessary to give us a fair shake, as it were, and therefore, remedy the situation.

In this scenario, deception would have to be employed to a great degree. There are good reasons for this. First, unlike a mother (or parent if you wish) and her child in the literal sense, the aliens (our parent or parents, figuratively speaking) cannot literally be with their child (us being the child or children, figuratively speaking) in an overt fashion. That is, a mother does not hide from her child. The child sees the mother, and the mother sees the child during feeding, clothing, and bathing, for example, and both consciously are aware of each other's presence in real time and develop a close bond in which all the necessary properties of friendship would emerge. There is no deception between the mother and child in this sense. Whereas, the aliens, as our caretaker (the parent), would have to remain a complete mystery to us (us being the child) as to their very existence and their purpose as much as possible. This would be especially true if they were our caretaker at our current evolutionary stage, despite of the fact that they would genuinely care for

[34] Deception here is applied in the collective sense, because this work is not exploring friendship between two individuals but two civilizations. Hence, no harm in this context is no harm as it relates to our species as a whole. The consequence of this distinction will be illuminated in due course of this work.

our species in this hypothetical scenario. There are two reasons for this. First, given (P7) and all that it entails, it would be wise for the preservation of the stability and safety of their civilization, if the aliens remained a mystery to us. Second, it is reasonable to assume that the aliens would be aware of our psychological and philosophical lack of maturity as a civilization. Moreover, they would realize that significant amount of humans would not be able to psychologically reconcile the reality they believed to be true versus the actual reality the aliens would be presenting, i.e., that they have been and continue to manipulate our evolutionary progress without our permission, for example.

How is it that the aliens would be able to covertly be our caretaker? That is, how would the aliens be our caretaker without us realizing it? Assuming (P1) and (P2) are true (true in the extreme sense, of course, as established in the beginning of this work), then it would be feasible for the aliens to capture us without our knowledge, then take DNA samples, alter our biology in some manner, monitor our overall development, and then return us back. In other words, the aliens would do what needs to be done in order to manipulate our development through covert means without our conscious knowledge and consent. It is reasonable to assume that their technology would seem like magic compared to ours, and they would be able to come in and out of our atmosphere in crafts with ease and still remain a mystery. They would be able to enter our homes in a "magical" fashion and take individual humans and do what needs to be done as described above. To ensure the covert mission is successful, it is reasonable to assume that the aliens would remain as silent as possible during the capture, and say only things that would need to be said to those captured so as not to reveal too much, and even provide conflicting information to the captured, i.e., providing confusing and false information, and also perhaps alter or erase the memories of the captured in some fashion, just like our intelligence community has done to their captured in the past. Providing little information and conflicting information to human specimens ensures that even if there were to be some residual memories, the collective memories of human specimens would be vague and contradictory. This would be magnified, given that the brain can subconsciously confabulate memories. As illuminated earlier, given what (P7) alone suggests, the aliens would be in great danger, if they were to provide us with their scientific and technological knowledge. Accordingly, the aliens would have good motivation with the utilization of deception along with the tactics employed as just elaborated. Of course, the duration of their mission would be long, complex, and arduous in order to accomplish their goal of helping us in this manner. However, it would be a manageable task, given the advanced technological status of the aliens.

Is there precedence for considering this kind scenario in some fashion? The short answer is yes. The method and tactics employed in the scenario just presented are not unlike what we currently employ to various kinds of species of animals that we are caretakers of, like the whooping crane, for example.

The mission of the Patuxent Wildlife Research Center in Maryland is to "excel in wildlife and natural resource science, providing the information needed to better manage the nation's biological resources" (*Patuxent*). One of their goals include the prevention of the extinction of whooping cranes. There are other research centers, zoos, and foundations, which are involved in this endeavor as well. Given the increasing loss of habitat of whooping cranes and unregulated hunting, the cranes were on the brink of extinction by mid-twentieth century. In 1942 there were only 16 living whooping cranes known to be alive in the wild. While they are still a threatened species, there is definite progress. Currently there are over 500 whooping cranes ("Flight"; *Patuxent*).

In raising the crane chicks deception in the extreme is employed. In handling and feeding the crane chicks the researchers dress in white like the adult crane and have their face covered so as to disguise the human form. They feed the chicks in total silence, so no human voice is heard, and do not display any emotions, despite what they might feel. Finally, they utilize a puppet head resembling that of an adult whooping crane so that the chick crane effectively learns from an adult crane on how to feed. This strategy is employed so that the crane chicks are not imprinted with the human form; so that they are not socialized by humans, but effectively socialized by another crane. When possible, researchers would concomitantly put a "live whooper adult as a next door neighbor," because that adult would aid the crane chick in learning to be a whooper himself (*Patuxent*). This, of course, aids in the proper imprinting of the crane chick. This kind of deception continues until they are full grown so that ultimately the scientists not only teach the cranes how to feed themselves, but also teach them how to fly and migrate, i.e., become a whooping crane that can be let out into the wild, live in the wilderness, and propagate itself properly.

Throughout their growth, deception in the extreme is employed in helping the cranes as a species not just to survive, but also to thrive ("Flight"; *Patuxent*; "Raising Crane").

Consequently, it is not too much of a stretch of one's imagination to envision that extremely deceptive tactics could feasibly be employed by aliens with respect to being our caretaker. Admittedly, the real life example of human/whooping crane deception is not a one-to-one relation to the alien/human deception example, but some of the key elements are there, and the human/whooping crane deception example is instrumental in conveying a general point across effectively. That is, given the whooping crane example, there is precedence for the notion that a civilization could employ deceptive behavior towards another civilization that they deeply care about in a somewhat similar sort of fashion, qualitatively speaking, as just discussed. Accordingly, in this context whatever the aliens might do, they would necessarily employ extreme deceptive strategies in dealing with us, even if they deeply cared about us as a species.

There are two other considerations. First, while the approach with whooping cranes seems to employ much deception, I do not mean to suggest that such a level of deception is needed with the caretaking of other animals. Indeed, the approach we enact on our caretaking of various species of animals depends on the species; different approaches for different species, and therefore different methods and degrees of deception. Second, being a caretaker of other species of animals also tacitly suggests that the population of the species is to be controlled. The caring of whooping cranes just discussed implicitly suggests this. As mentioned earlier, the population of whooping cranes dwindled almost to the point of extinction due to increasing loss of habitat and unregulated hunting. So, this project is a positive one in that the goal was not only to increase the population of whooping cranes but within the context *letting* them live and thrive in the wild. We do these sorts of things with other animals as well. A thing to consider when controlling the population of one species of animal is that it has a causal effect on the population of other animals we care about deeply and the ecosystem of a particular area as well. We control various populations of animals by means of various chemicals. And we do this when we conduct warfare with the lower life forms on our planet in protecting our loved ones and home. We do this with insects and bugs all of the time, for example. Ever use a can of spray to kill hundreds or even thousands of ants that have infested your kitchen? There are various companies that make a living out of managing the population of insects and bugs (and even certain mammals like rats etc.) within all of our homesteads. We do not mind if they are outside. However, once there are too many of them, and they start eating all of our fruit in the fruit trees or they start nesting in our yards or inside our homes, e.g., termites, then we wage chemical warfare on them; not to kill every living ant or termite on the planet but enough to keep them out of our dwelling. So it is in this basic and brutal sort of way that we manage the population of certain kinds of animals as a matter of course. However, what I am concerned with here are animals that we, as a whole, deeply care about, where politics and emotions are high with respect to certain kinds of animals and how we manage them. What if a population of an animal we care deeply about was indeed overpopulating itself to the point of not only endangering its own environment but the environment of other animals and the surrounding ecosystem? Would we interfere and manage the population of that animal and even resort to culling? Is there precedence in actually reducing the population of an animal that we deeply care about, by culling, or by use of other means, due to the reasons just mentioned? The short answer is yes. I shall present one such example, utilizing the elephant to illuminate this point.[35]

Managing the population of elephants is of much concern in various nations in Africa, e.g., South Africa, Kenya, Zimbabwe, Angola, Mozambique, and Namibia. Certain African nations are reconsidering the ban on culling elephants, given the rise in their population and how the elephants affect the environment in the regions they inhabit. For example, in Zimbabwe since 1980 the population of elephants has risen from 46,000 to over 100,000 as of today. As a result in this increase, the government of Zimbabwe claims that the population increase of elephants is twice as many as the land can support. In South Africa's Kruger National Park, culling was implemented from 1967 to 1995 with result of reducing the population of elephants by 14, 562. However, since South Africa's banning of culling in 1995 the population has risen from 8,000 elephants to about 13,000. Moreover, each elephant consumes about 400 pounds of food daily.

---

[35] I am not suggesting that all humans care deeply about elephants (or even whooping cranes) but only that there are enough humans that do. Again, the relationship with these kinds of examples are not wholly one-to-one with respect to how aliens might treat us, but certain key aspects can be easily illuminated by the use of these kinds of comparisons in getting a proper grip on what I am trying to convey.

As a result of the increase in the elephant population, along with their appetite, the elephants in Kruger National Park are transforming the landscape, by "tearing through vegetation, pulling down or uprooting trees and stripping them of bark." The result is the conversion of some of the landscape from wooded savanna to scrub grassland, which provides habitat for animals that graze like zebras and, also, provides "nesting places for eagles and other birds." Of course, there are other methods that were employed, i.e., contraception, translocation, a multipronged approach, and the free-range approach. However, in the end they are not as viable as one would intuitively imagine. That is to say, there are challenges with these kinds of approaches. For example, with contraception, the cost is "more than $150 per elephant and must be done repeatedly." And with large parks like the Kruger National Park, it is economically not a viable option. Although, in other protected areas such as the Makalali Private Game Reserve, which may only accommodate smaller population of elephants of about 70, the approach has had success. Nonetheless, this has changed since 1979 when about 30 reserves that were considered small have accepted elephants from the Kruger National Park and "are now struggling to manage their own population of elephants as well." Translocation is also costly with "few places left that are large enough to accommodate an influx of elephants." The bottom line is that given the difficult challenges that arise out of these other methods of controlling the population of elephants, some specialists are considering culling to be implemented once again in South Africa at some point. In Zimbabwe, the government has already allowed the killing of 150 elephants for meat (Lange).

Now, whether by culling or any other method, how would we respond, if we realized that there was an alien civilization that was controlling our population in the some fashion in a deceptive manner and without our permission? Would not they have the right to manage our population, given their superiority over us as we are to other animals? Would we care about their intentions and motivations, even if they were altruistic ones? I gather that, as a whole, it would not bode well with our civilization at all, putting it mildly.

Consequently, given, (P7) it is safe to assume, even if aliens deeply cared for us, because they saw our potential and wished to nourish it, they would still necessarily have to employ extremely deceptive methods with which to carry out their mission of helping us. Hence, it would not be wise for them to simply provide us their scientific and technological knowledge openly, given what (P7) alone suggests. For the aliens to simply provide us with knowledge about our planet and the universe is not enough. From their perspective, we simply are not mature enough psychologically or philosophically to handle the information could provide us or even to understand it, given that an alien civilization in this context would have a world view vastly different from ours. Also, it is important to note that the aliens would not take offense at our distaste and rage against their culling of our population, because I would gather that they would appreciate, all things being equal, that all life forms always try to assert themselves, and every intelligent civilization like to be in control of their own fate. However, the aliens would also be very cognizant that our rebellion against them would be on grander and more dangerous level, given that we do know how to split the atom after all, and as what (P7) suggests is clear. We are an incredibly violent species, and given (P5) and (P6) as well, the aliens have more than enough reason and motivation to implement deceptive behavior towards us, even if they truly cared for the well-being of our species and our planet.

Third Scenario of Deception – Negative Sense

The core properties of deception in the negative sense are thusly:

1. There would be interaction between aliens and us.
2. The aliens would be using us in some manner solely for their own benefit.
3. Given what (P5), (P6), and (P7) collectively suggest, the aliens would be motivated in practicing deception to the best of their ability; remain completely mysterious, or hidden from us as to their presence and their intentions.

This kind of interaction should be construed as unfriendly, because the aliens would be using us in some manner without our knowledge and solely for their own benefit. However, there would not necessarily be any ill feelings from the aliens towards us as such. That is to say, there would not necessarily be any hateful or malicious feelings from the aliens towards us. Their attitude and actions towards us would be from a purely pragmatic standpoint. Also, one thing to consider is *just because the aliens could be unfriendly towards us does not necessarily entail our immediate destruction or any destruction of our existence as a species*. I label this supporting proposition for (EP7) within the context of this third scenario of deception in the negative sense (italicized portion) as (EP7A).

Elaboration of (EP7A)

Consider the philosophical text known as *The Art of War* attributed to Sun Tzu (544-496) BCE, who was a Chinese philosopher and military strategist.[36] Sun Tzu employed a Taoist approach to warfare in his work. For the Taoists, "life is a complex of interacting forces"[37] in which "intuitive cooperation with the natural order,"[38] referred to as the *wu-wei* or *non-action*, should be the approach to life. This also incorporates the view of opposites (oppositional dualism). That is to say, when starting something, then many times one should start with the opposite. This quote from a Taoist text known as the *Tao Te Ching*, attributed to Lao Tzu, exemplifies a typical Taoist approach to something, and in this instance, war:[39]

> The best warrior is never aggressive.
> The best fighter is never angry.
> The best tactician does not engage the enemy.
> The best utilizer of people's talents places himself below them.
>
> This is called the virtue of non-contention.
> It is called the ability to engage people's talents.
> It is called the ultimate in merging with Heaven. (Tzu, Lao Chapter 68)

Now consider these quotes from the *Art of War* by Sun Tzu:

> A military operation involves deception. Even though you are competent appear to be incompetent. Though effective appear to be ineffective. (49)

---

[36] I wrote *attributed*, because scholars do not know how much of the work Sun Tzu himself wrote, and thus, believe that it is a composite work of various thinkers that have contributed to it as well. Accordingly, for scholars the name Sun Tzu is more of an indexical name as the "author" for the complete work of *The Art of War*.

[37] From Thomas Cleary, within the *Translator's Preface*, page VIII. Please refer to *Tzu, Sun* in the *Works Cited.*

[38] From Yi-Ping Ong, within the *Introduction*, page XVIII. Please refer to *Tzu, Lao* in the *Works Cited.*

[39] I wrote *attributed*, because most scholars view Lao Tzu as a mythical person created by the Taoists. Most scholars view the *Tao Te Ching* as a composite work incorporating many Taoist writers. Accordingly, the name Lao Tzu is more of an indexical name as the "author" for the complete work of the *Tao Te Ching*.

> Be extremely subtle, even to the point of formlessness. Be extremely mysterious, even to the point of soundlessness. Thereby you can be the director of the opponent's fate. (104)

> The general rule of the military is that it is better to keep a nation intact than to destroy it. It is better to keep an army intact than to destroy it, better to keep a division intact than to destroy it, better to keep a battalion intact than to destroy it, better to keep a unit intact than to destroy it. (66)

> Therefore one who is good at martial arts overcomes others' forces without battle, conquers others' cities without siege.... (72)

Throughout history, there were nations that expanded their borders and would utilize non-violent methods or bare minimum violence for purely practical reasons that were reflective of the kinds of things asserted in *The Art of War*. That is to say, why invade a civilization violently when it could be done nonviolently with the least resistance? Hitler, for example, initially expanded the borders of Germany without any bloodshed. With a bit of political maneuvering, Hitler annexed Austria in 1938 and Sudetenland (i.e.., northern, southwest, and western parts of then Czechoslovakia). In 2014, Russia made a land grab in a certain region of Ukraine, i.e., Crimea with a population of about 2 million, which resulted in only six casualties. I am sure a proper military historian could provide even more examples that supports this thesis. Indeed, the *Art of War* by Sun Tzu is still on the reading list at the West Point Military Academy for *The Officer's Professional Reading Guide Top 100 List* ("The Officer's"). The point is that no matter the motivation of the aggressor, there is precedence for thinking and enacting views that are reflective of Sun Tzu's philosophical work on warfare, even in our own time. Consequently, even in warfare sometimes the best strategy is not to employ overt obvious sort of violence at the start or, perhaps, ever.

Do we not wage war or interact in obtrusive ways with other animals on a daily basis as a matter of course and employ deception? We do not negotiate with these nonhuman lifeforms as to how we can work together. We just implement force or deception in the extreme motivated by our benefits. Indeed, many humans have such a superiority complex that many do not even consider us as animals, contrary to all of the scientific evidence suggesting otherwise, which helps rationalize the aggression towards other nonhuman civilizations on our planet that perhaps should be treated better, e.g., dolphins. Nevertheless, as mentioned earlier, we wage war against smaller animals such as bugs with chemical sprays that kill ants as a matter of course, for example, when we feel they are invading our homes.[40] In fact, we do kill ants and other insects in the millions on a daily basis throughout our own nation (i.e., the U.S.). In addition, most of us are not motivated by hate for these bugs, as such, but only out of necessity of keeping our homes in order. But what of deception? We kill slugs and snails by employing deception. That is, we put what they perceive as food in our gardens, and the slugs and snails eat this food (bait), and then they eventually die. Many times we do not necessarily kill other animals, but, instead, utilize various animals for all sorts of ways solely for our own benefit in which deception is employed. Sometimes deception does not even have to be employed, because the animal is not intelligent enough to even perceive its own situation. At any rate, I shall present pigeons, cattle, dogs, and silkworms as examples of animals that we do not necessarily hate, but, nevertheless, we utilize them in deceptive ways solely for our own benefit, which supports the main thesis of this section that killing or blatantly taking over another civilization in a violent manner is not necessarily the best or productive strategy of dealing with a civilization.

Homing pigeons have been especially useful for us. We have selectively bred these homing pigeons because of their natural instinct of being able to navigate back home over vast distances. We utilized these pigeons in WWII, for example, for the purpose of passing on messages. Hence, we have risked the pigeons' lives solely for our own benefit during the time of war without the pigeons realizing that we put them in danger for our purpose only.

We capture cattle and use them for all sorts of things solely for our own benefit, e.g., food, clothes, and sports entertainment. Indeed, with many cows we go to the same ones over and over and over again throughout their lives so as to retrieve a great source of food: milk. For entertainment, we use bulls in rodeos. We keep them alive and use deception and get them to do what we want them to do. Bullfighting is still popular in certain parts of the world. We raise the bull to its

[40] I use the term bug(s) in the loose sense as a catchall term referencing any tiny creature with legs, e.g., ants, spiders, beetles, or bees etc.

full maturity. We keep it happy and feed it well. The bull has no idea that all this is part of a plan for its eventual slaughter for food within the context of entertaining us.

We have selectively bred wolves into domesticated dogs for a variety of reasons, and did so without negotiation, but, instead, with manipulation and deception. We, of course, have dogs as pets, but we use them in other ways as well. For example, it is common practice for various sorts of businesses in utilizing dogs as protection against thieves by appealing to their natural instinct for hunting and guarding, then manipulate their predilections solely for our benefit. We use dogs to detect cancer in people, given that dogs have a strong sense of smell. We put dogs in danger in wars for our purposes only. Dogs have no awareness of the fact that we bred them a certain way for our own goals and whims.

We harvest silk from silkworms in which the silkworms are not immediately killed. When they hatch, we feed them all the mulberry leaves they desire and take good care of them. Once they are in the pupa stage and are in their cocoon, we then boil the cocoons thus killing the pupae before they can transform into a moth so that we may properly harvest the silk from the unbroken cocoons ("Silk").

Do we use other animals to control or manage other animals?[41] Yes, we do. I shall present two brief examples with dogs and ladybugs as illustrating this point.

We use dogs to herd other animals like sheep and cows for us. Of course, we appeal to their natural instinct of play, herding, and hunting and manipulate those predilections solely for our purposes unbeknownst to the dogs themselves. The dogs are not harmed of course, but we have bred certain kinds of dogs as herding dogs for our bidding. In this way we have altered their genes and their biology solely for our own benefit. Is not this deception in the negative sense? Would we want aliens to do this to us in some fashion?

We utilize ladybugs for managing the population of certain bugs we do not like. That is to say, we use them for killing various small bugs like aphids and spider mites, because we do not want

[41] This would include killing as well.

these animals on our plants, e.g., milkweed and roses. One can go to a local garden center and purchase ladybugs for said purpose. These bugs are a commodity solely for our desires. Some inadvertently are killed due to the process of shipping, but as a species, they are protected and harvested by various industries so that they can make a profit off of selling these animals for garden enthusiasts who seek a more organic or sustainable way of managing (killing) other undesirable animals we commonly refer to as pests. Hence, as we breed ladybugs, we also make sure their population does not wane, yet, it is accepted that some will parish during this whole process.

The various kinds of animals that we utilize for our own benefit are seemingly endless and the reasons for utilizing them is seemingly endless as well, and I am sure upon reflection the reader can come up with plenty more examples that I have not mentioned, which supports the thesis within this section of this work. I have merely presented what I consider to be some obvious and interesting well known examples of animals we utilize solely for our own benefit in which the animals as a species are not exterminated but utilized in a deceptive manner in varying degrees as necessary. Note that within this context it is acceptable that a certain percentage of individual animals of said species would perish, so long as the species as whole is preserved, so that we may utilize them over and over again.

Even in the non-human animal kingdom on our planet, not all animals immediately kill the other, but instead keep the other alive to serve a certain purpose, and once the purpose is fulfilled, then the victim would be killed, if necessary. I shall present three such examples; the first two in which an animal uses a form of mind control on another animal, and the final example in which a brood parasitic approach is implemented.

Some animals employ using mind control methods on other animals for solely their own purpose, then dispose of them after the purpose is fulfilled. Let us take the female Costa Rican wasp,[42] for example. First, the wasp lays its egg in the abdomen of the orb spider.[43] Then, the wasp larva lives inside the orb spider for a few weeks. Afterwards, the wasp larva injects a chemical into the

---

[42] *Hymenoepimecis argyraphaga*
[43] *Plesiometa argyra*

spider that will end up controlling the spider's mind. That is, due to the chemical injection from the larva, the spider goes and constructs a special web that would be able to support the cocoon that the wasp larva would build. Finally, unfortunately for the spider, after the special web is constructed, the larva eats its way out of the spider, thus, killing it, before building its own cocoon. Note that the adult wasp does not kill the spider, it is the wasp larva that actually kills the spider. But only once the spider's purpose is fulfilled, i.e., nourish the larva and construct a special web for the larva (Bates).

There is another wasp that practices mind control in which the victim can go unharmed. The victim is used, then goes on its merry way living out its life as if nothing occurred. It is a wasp known as the Dinocampus coccinellae that lays its egg inside the ladybug by virtue of its stinger. The wasp larva grows inside the ladybug and in about three weeks as its transitioning into an adult, it "spins its cocoon between the ladybug's legs and starts the transformation" (Wilcox). One would think that at this point the ladybug would surely attack the larva, but instead, "the ladybug stays with the cocoon, protecting it until the wasp hatches" (Wilcox). What occurs next is even more bizarre. That is, about 25% of the time the ladybugs survive this ordeal unharmed, and simply return to living its life as if nothing occurred (Wilcox; "Dinocampus coccinellae").

Some animals are brood parasites. A brood parasite utilizes deceptive means in order to get another animal of another species to raise its young. The cuckoo bird employs such a strategy by laying her egg in a Reed Warbler's nest when it is not around. Once the baby Cuckoo bird hatches it also gets rid of its competition by removing the other eggs in the Reed Warbler's nest. This strategy ensures that it will get fed well by its newly adopted mother, the Reed Warbler (Dobkin et al.).

It is evident that deception in the negative sense is, as presented here in this section of this work, a useful strategy for survival in nature. Deception within this context minimizes work and danger for the one who applies deception. After all, it is better to *manipulate* your opponent than outright *confront* your opponent. This is because, the former approach minimizes danger for the aggressor (the deceiver), and puts the workload on the opponent (the deceived). In the examples presented above, deception involved utilizing what was already there. Metaphorically speaking, it is a way of 'going with the flow,' as it were; a manner of achieving a goal with least resistance. Whether it is us deceiving other humans, or us deceiving other animals, or other animals doing the same to other animals, in every case presented, it seems to involve taking advantage of another's natural disposition. For example, when we use dogs to our advantage, as presented above, we merely appeal to the dog's natural inclination to hunt and guard. The wasp takes advantage of the spider's natural inclination to build webs, and so on. In the case of dogs, we also breed them a certain way so as to enhance certain behaviors or other qualities we desire from them. Deception within this context is not only useful in getting labor and resources, but, as illuminated by Sun Tzu, deception in the negative sense is useful in conquering an enemy. This is evident not just with how we behave but also evident with the behavior of many other various life forms on our planet. *Please note, the examples I provided were examples in which hate was not a motivator, yet, clearly the actions could be considered unfriendly to the one on the receiving end.*

The obvious question is what would aliens want from us that would encourage them to behave deceptively in the negative sense? One reason could be that aliens might want or need something from our biology. Perhaps, there could be something that aliens would harvest from us as a resource as we do with cattle and other animals over and over again for some medical or other biological purpose we have yet to conceive of. The second reason for employing this kind of deception is perhaps the aliens would like to inhabit our planet. Biological incompatibility among species and the environment is the issue here. Let me elaborate.

It could be the case that the aliens desire or need a new place to live and our biology is conducive for that end. If an alien civilization were to live here, it would be reasonable to assume, even if they were a carbon based life form, that they would come from a planet with very different oxygen, gravity, and ecological system. Accordingly, their biology might not be able to adapt to our gravity or oxygen in the long term, let alone the kinds of food their biology would allow them to digest nutrients properly. Even on our own planet what is dangerous for one species of mammals would not be dangerous for other mammals. For example, while chocolate is enjoyable for us, it can be toxic for cats. In the broader sense, mammals are diverse in their diets. For example, some mammals on our planet are herbivores (e.g., cattle) and others are carnivores (e.g., lions). Other mammals, like us, are omnivores. Even within the plant kingdom diets can vary. While most plants are not carnivorous, many are, e.g., Venus Flytrap among others. In fact, there are at least 720 species of plants that are carnivorous ("Grow Carnivorous Plants"). Furthermore, an alien's immune system could be quite different from ours and not adaptable to things that are harmless to us on our planet.

How would they live on Earth utilizing our biology without us noticing their presence? One approach could be that they would take over our planet and concurrently they would modify their biology so that they would aesthetically look like us. However, given the assertions in the previous paragraph, it would not be enough for the aliens to merely look like us. If they were to survive on our planet, then they would have to modify their internal organs as well. As established by (P3), it could very well be that an alien civilization would not have any qualms about altering their bodies as long as their consciousness resides in a body that allows them to live in such a way that their intelligence, cognitive abilities, and appropriate physical abilities remain intact.[44] As discussed within the context of (P3) the aliens would utilize science and technology as an ongoing process of constantly manipulating their own evolution as a way of advancing physical and mental capacities as needed. Hence, even though they might follow an overall biological model, their identity would not be solely regulated in thinking of themselves strictly in a biological sense but more in terms of a socio-political one. Accordingly, the aliens may simply view the body as a container of sorts. And this container would be modified in adapting the body's mental and physical capacities when necessary. Given the aforementioned as true, then by extension, it could be the case that an alien species could feasibly utilize a slow process of assimilation of another species instead of just outwardly destroying the threat—that would be us— and then simply inhabit the planet. It would be a practical process, because winning against an adversary without any violence and with that adversary not even being aware of a takeover reduces the danger to the deceiver, as indicated earlier.

The point is there are good examples in nature (that includes us) that supports the notion that to have unfriendly intentions or acting in an unfriendly manner towards another species or civilization does not necessarily entail immediate destruction or any destruction of that species or their

[44] By "appropriate physical abilities," I mean that the bodies they would inhabit allow them to live in the adopted environment properly.

civilization. The aliens would have their own time table as to when we would be destroyed or taken over, assuming they had this sort of plan in mind. Accordingly, if aliens were to interact with us in the negative sense, it is reasonable to assume that they too would interact with us in some bizarre deceptive fashion. If primitive animals can co-opt other animals to do their bidding by utilizing mind control strategies, then it stands to reason that animals that are considerably more advanced than us could do such things as well. That is, aliens might employ mind control strategies in ways we cannot even imagine. Also, the scenarios presented in this work do not necessarily suggest that the aliens would be evil as such. The aliens would be no more evil than any other animal asserting itself for the purpose of its own survival. Within this context, we would merely be a means to an end in some fashion. It is deception in the negative sense, only, because we would be on the receiving end of it. Hence, to assume that contact with a hostile alien civilization would necessarily entail our immediate or some sort of obvious destruction, or any destruction as a species, is shortsighted.

4. OTHER CONSIDERATIONS

Introduction

This chapter tends to some reasonable concerns one might have with respect to some of the views presented in this work.

Consideration 1: With Respect to Friendship

It could be asserted that friendship could develop between aliens and humans, if there were to be a mutual need. By working together, then perhaps a genuine friendship would emerge as a result. One reason for this line of thought could be that the aliens would gain a better appreciation of humans, and so they would develop genuine feelings of goodwill towards us. Is this plausible?

My Response

I appeal to Aristotle as the backbone of my reply, given that my sentiments agree with his reasoning. According to Aristotle, the kind of friendship that would be entertained would be one based on utility (1156^a10-13).[45] This kind of relationship is not sought out, unless the parties involved "are useful for each other" (1156^a28-29). In addition, Aristotle correctly warns us that within this context that the goodness between parties would be temporary, because "when the motive for friendship is done away, the friendship is dissolved" (1156^a23). Interestingly, given what has been stated, it is evident that this kind of friendship need not involve parties who even like each other, because the glue of friendship is solely based on utility. Aristotle states:

> But the friendship of utility is full of complaints; for as they use each other for their own interests they always want to get the better of the bargain, and think they have got less than they should, and blame their partners because they do not get all they want and deserve; and those who do well by others cannot help them as much as those whom they benefit want. (1162^b17-21)

By extension this can be applied with respect to friendship among nations. We only have to look at our own history to confirm this. There are many examples throughout history, which support

[45] All of the parenthetical references and quotes of Aristotle are from his Nicomachean Ethics.

the notion that relationships between nations that were purely based on utility, ended disastrously for at least for one of the parties involved. I shall present two such cases.

During WWII the Soviets collaborated with the Germans. Germany and the Soviet Union signed a non-aggression pact in 1939, referred to by various names, e.g., The Molotov–Ribbentrop Pact and Nazi–Soviet Pact. Ostensibly, it simply asserted that both parties agreed not to attack each other. That is to say, while the Nazis were conquering Western Europe, the USSR promised to turn a blind eye as to the German activities. In turn, while the Soviets were conquering nations on their end, the Nazis would turn a blind eye as well as to the Soviet activities. However, this pact also included a secret agreement that addressed how certain parts of Eastern Europe would be carved up between the USSR and Germany. By 1941 Germany would break the pact with the invasion of the USSR, thus forcing the Soviets to militarily engage Germany. This friendship based on utility between the Germans and the Soviets is an example of the instability and danger of a friendship solely based on utility.

Remember the old adage, "The enemy of my enemy is my friend"? Later on, during WWII, the US collaborated with the Soviets. That is, the US and the USSR developed a friendship based on utility to fight Germany despite the fact that the US government was well aware of the horror and extreme oppression and violation of human rights of Stalin's regime, where millions were murdered preceding the war and during the war. When this friendship of utility wore off, the USSR and the United States eventually and predictably turned on each other. A cold war arose, and in the end the relationship almost ended in to a hot war, i.e., The Cuban Missile Crises.

The implicit message is clear. That is, a friendship based on utility between the human civilization and an alien one, although possible, is not truly a desirable one, because it would probably end up being disastrous for us due to what the SP collectively suggest about extraterrestrials as defined within the context of this work. Furthermore, one must wonder, why would an alien civilization as described in this work desire a friendship with humans let alone one based on utility? There is only one possible condition for them to consider such a friendship. If an alien civilization were to decide to build a friendship solely based on utility, then there must be some pragmatic reason and ulterior motive for this decision, because given what (P1), (P2), (P5), (P6), and

(P7) collectively suggest, there would seem to be no reason why aliens would need to come to any sort of agreement with us; they can just take what they want through deceptive means. It is reasonable to assume that if an alien civilization were to accept a friendship based on utility with humans, then it would be only because they need X, and X would be, for whatever reason, easier to attain within the context of a friendship based on utility. If the aliens were to try to establish a friendship with humans, then it is reasonable to assume that there would be an ulterior motive for doing so, especially given (P7) and (P7A). Once their goal would be met, the friendship would dissolve, and we would then most likely end up in a dangerous situation in which we would be on the losing side, given their technological superiority alone.

Perhaps a caveat to my argument against friendship based on utility with aliens is that this kind of relationship could be a useful strategy for us in buying time so that more data could be acquired about the extraterrestrials. As a result, the friendship based on utility has its place. Still, this would be a dangerous game. Yes, I said *game,* because there is a game like quality to this kind of relationship. Unlike a genuine friendship which is not a game, a friendship based on utility does not have the property of goodwill, which is a necessary property for genuine friendship. Also, one should assume that the aliens would be well acquainted with deception and all of its forms, especially ones that would be related to the friendship based on utility. Hence, in the end, again, this kind of friendship between aliens and humans would most likely put us on the losing side of the friendship, especially once this relationship dissolves.

Consideration 2: With Respect to the Rudnyk Scale

I shall present three scenarios in the form of questions with elaborations where needed that present challenges to the Rudnyk scale and, then present my response. I shall also address an additional concern. Finally, I shall then present my final thoughts with respect to the Rudnyk scale.

Scenario 1

What if a Type-I civilization had a naturally occurring wormhole within their solar system that was close to their planet, which enabled them to travel to other solar systems?

Scenario 2

What if some extraordinary event occurred to a Type-I civilization, which gave that civilization access to technology that in turn accelerated their own technological development for space travel to a Type-III status? An extraordinary event that would allow a Type-I civilization to obtain a Type-III craft [46] could take on various forms. Here are just a few examples:

- Discovering a crashed alien craft due to a malfunction of the craft.
- Discovering a crashed alien craft due to a meteorological event that caused the crash.
- Discovering an alien craft due to a mid-air collision with one.
- Discovering an alien craft due to shooting one down.

Scenario 3

What if a Type-I civilization had a naturally occurring element that was present on their planet, which they in turn discovered that would be rare or nonexistent on our planet or even our solar system, but greatly aided their technological development for interstellar travel?

The Main Question with Respect to the Three Scenarios

Would the kind of scenarios presented above be a counter to the view that only a Type-III civilization could be an interstellar civilization and puts into question the validity of the thesis the Rudnyk scale puts forth as a whole?

My Response

[46] The expression "Type-III craft" or "Type-I technology" is a shorthand for "a craft that is (or would be) reflective of a Type-I civilization." It is in like manner with the expressions of Type-I or Type-II craft or technology and so on.

My overall rejoinder with respect to all three scenarios is that the Rudnyk scale implicitly operates from the "all things being equal" clause. That is to say, this work deals with a notion that a civilization that would have evolved a certain way would have done so under equal[47] circumstances with that of any other civilization. This kind of approach by appealing to the "all things being equal" clause is a common approach philosophers and scientists alike utilize when dealing with a thought experiment, an actual experiment, or conceptual analysis of something so as to get a basic understanding of X. For example, if scientists were to conduct an experiment with how humans react to a hearing a loud sound, then the experiments conducted should naturally only include those humans who could actually hear and have a normal hearing range; hence, effectively implementing a condition of the "all things being equal" clause, in taking affect, even though there are plenty of humans that cannot hear, or cannot hear well, or have varied hearing capabilities.

With respect to the first scenario, I would assert that a Type-I civilization that would be reliant on a wormhole for interstellar travel would not truly be a Type-III civilization, because it would have utilized the wormhole and not its own technology to get from point A to point B. And once this Type-I craft would emerge from the wormhole, it would do so most likely with its chemical propulsion, which would be a dead giveaway that the craft would not truly be reflective of a true Type-III civilization. It would be somewhat analogous to finding a shortcut when hiking. Utilizing a shortcut can be a smart strategy in getting from one point to another quickly when hiking, but this would not be reflective of any sort of technology that enabled one to travel faster and farther as such.

With respect to the second scenario, I am hesitant with the notion that a Type-I civilization would be able to figure out how to use such advanced and different technology that would be reflective of a Type-III civilization much less being able to back engineer such a craft. It would be somewhat analogous of civilization from a thousand years ago (a Type-0 civilization) in figuring out how to properly launch our space shuttle craft much less back engineer it and replicate it. In fact, it would be dangerous for such a primitive society to take apart such a craft, given their lack of knowledge of the dangers of electricity and certain chemicals that would be present within the

[47] Equal in a qualitative way.

craft. Indeed, a Type-0 civilization would most likely perceive Type-I technology as something from the gods. It would just be perceived as a mystery, or possibly, if something dangerous occurred due to an accident, as suggested earlier within this paragraph, the Type-0 civilization might end up avoiding it.

On the other hand, there is a great deal of difference between a Type-I and a Type-0 civilization in important ways. First, scientists from a Type-I civilization would be less prone with associating superstitious beliefs to extremely advanced technology obtained in one manner or another than a Type-0 civilization. Second, while a Type-I civilization might not necessarily be able to back engineer a Type-III craft that would be in their possession, unlike a Type-0 civilization, a Type-I civilization would have a great deal of knowledge of physics (like we do) with which they could theorize about how the craft moved around and work on creating their own, which could end up being different from the alien craft but still be able to do many of the kinds of things that a Type-III craft could accomplish. It is important to note that there could plausibly still be unforeseen dangers for a Type-I civilization of trying to back engineer a Type-III craft, given that a Type-III craft would necessarily be different in the extreme as (P2) asserts. I do concede that the second scenario would provide conditions that could possibly encourage a faster pace of technological development for a Type-I civilization, simply, because thinking that *"something was considered impossible is actually possible"* itself motivates change for the development of more advanced technology. As discussed within the context of the elaboration of (P2), the first necessary step for a civilization in developing radically advanced technology is within the context of a paradigm shift in thinking, i.e., a new world view. As mentioned early in this work, discovering something is one thing, but having the mindset to endeavor to truly appreciate and understand what one has in one's possession requires a certain mindset or world view, not just from the discoverer, but also from the respective society and political system as a whole for a discovery to properly take root within that civilization.

While the second scenario would provide the conditions for quicker development of radically advanced technology for a Type-I civilization, it is only a civilization that would have the predisposition of being open to new ways of understanding things in the first place that would allow for quicker development of advanced technology. So it is not knowledge that would be simply given in this scenario, but knowledge that would still have to be earned due to the fact that the Type-I civilization would still be relegated in figuring out for themselves with how the craft actually worked.

With respect to the third scenario, I appeal to some of the arguments just provided in the previous two paragraphs. While the third scenario would provide the conditions for quicker development of radically advanced technology for a Type-I civilization, it is only a civilization that would have the predisposition of being open to new ways of understanding things in the first place that would allow for quicker development of advanced technology. So it is not knowledge that would be simply given in this scenario, but knowledge that would still have to be earned due to the fact that the Type-I civilization would still be relegated in figuring out for themselves with how to utilize the element discovered on their planet for transportation. And to do so that society's whole world view would have to change, because discovering something is one thing, but to have the mindset in endeavoring to truly appreciate and understand what one has in one's possession requires a certain mindset or world view, not just from the discoverer, but also from the respective society and political system as a whole for a discovery to properly take root within that civilization. So, a Type-I civilization that would have eventually learned how to utilize an element that would naturally be found on their planet that could accelerate the rate of that society's traveling technology could still properly be referred to as a Type-III civilization as described in the Rudnyk scale.

Discoveries that are profound are so because they just do not advance a civilization's technology but also advance a civilization as a whole in all areas of that society. If something occurs prematurely, then danger in the extreme would be present, and then nature would balance this out. This is the way of nature; something that is not truly ready for X will in the end not have X. It would be somewhat analogous to a chimp that would have a loaded gun when it should not. The consequences are obvious. The chimp would not just unknowingly be a danger to itself but to the rest of its group. Nature has a way of balancing itself out. A Type-I civilization will never attain a Type-II status or a Type-III status until it is truly ready. A civilization that cheats will ultimately self-destruct.

Additional Concern in the Form of a Question

Why does the Rudnyk scale address the aspect of moving around (traveling) in terms of a vehicle? Perhaps a civilization could be so advanced that they would have created a device that can create a gateway that would let individuals to merely walk through a "gate" to any planet within the galaxy. Hence, a vehicle for traveling to other planets within the galaxy would not be needed.

My Response

With respect to the this scenario, still, all things being equal, this kind of traveling or moving around would still be reflective of an extremely advanced society, i.e., at the very least a Type-III. I just utilized the notion of a vehicle for clarity and simplicity as a strategy that provides support for the overall thesis of the Rudnyk scale, i.e., the Rudnyk scale lets one ascertain the intelligence of a civilization by simply observing how that civilization moves around.

Final Thoughts about the Rudnyk Scale

While I appreciate some of the concerns one might have about the Rudnyk scale, in the end, the concerns are not strong enough to warrant the abandonment of it as such. In addition, the Rudnyk scale is to be understood as a guide, not as absolute law, because I appreciate the fact that nature can be quite messy and does not necessarily put things in a nice neat box. Moreover, with respect to the kind of society associated with each civilization type, this, too, is to be understood as a guide; educated, thoughtful, and reasonable assumptions utilizing human history and life on Earth as reasonable indicators.

Consideration 3: With Respect to Aliens as Biological Entities

Some might ask why I assume a biological model of an extraterrestrial within this work, given that an alien civilization could, in principle, be an AI robotic/cybernetic type of species.

My Response

This concern directly relates to the second condition of this work from the Introduction and (P3) from Chapter 2. My first response is that even if the aliens were a robotic or cybernetic sort of civilization that the SP and the EP would still apply. There is no reason to think otherwise. The obvious rejoinder to my response is then why would I not just include this sort of consideration in my definition of the term *alien* within the context of the second condition that I put forth in the Introduction in the first place. It is a reasonable concern and my additional response is thusly:

As stated in Chapter 2, I appealed to Thomas Kuhn and argued for the view that scientific and technological progress can only be made once a civilization is able to have a different world view from its previous one, i.e., a paradigm shift. However, the strategy of how to implement this is a different issue. With respect to why I assert that a biological model of an extraterrestrial is the most plausible scenario as opposed to a robotic/cybernetic model stems from the notion of the path of least resistance. Let me elaborate.

If any intelligent species were to continue their evolution, they would do so in ways that would be reflective of the Universe's natural inclinations. What I mean by *natural* is that something which occurs or exists without conscious intent and effort; what nature is predisposed in doing even without conscious intervention. I realize that a rejoinder to this assertion in a form of a question could be, "Could not one assert that conscious beings such as us or any other intelligent conscious beings are *of* the universe and not something separate from it, and therefore, even something that is created by something conscious can be considered natural?" This, of course, is a reasonable question. My response is yes that in the broad sense everything that occurs in nature is "natural" in that it occurs in the universe, but what occurs in the universe, i.e., with conscious effort and intent, is subservient to pre-existing laws and properties of the universe, and that is what I mean by *natural* within this context. Moreover, the Universe is not predisposed in naturally creating robots or AI life. That is why the "A" in AI is "artificial." Without conscious intervention it is impossible for robots to come into existence. Without conscious intent and effort it is possible for life to come into existence into the material world. Indeed, it has for billions of lifeforms on this planet alone. Biological life occurs *naturally*; by default. The properties of nature are set up in such a way that when life appears through matter, it is biological. The brain is a biological object and the body, which is in extension of the brain, is also a biological object and itself is made up of many other biological objects, e.g., skin, eyes, ears, nose, tongue, legs, arms, heart, lungs etc. Furthermore, the body and its associated biological objects naturally have, all things being equal, various properties of the senses and other properties that living things have in general, e.g., touch, smell, taste, hearing, vision, growth, healing, etc.

Accordingly, if beings from an intelligent and technologically advanced civilization, i.e., a Type-III civilization, were to modify their own bodies, they would already know that the best strategy that they should adopt is the one that would be in the context of working with materials that already have nature's pre-existing properties, and simply continue from there, as opposed to starting from scratch and going uphill, as it were, by working with materials that have none of the aforementioned pre-existing key properties of nature.

5. CONCLUSIONS AND FINAL THOUGHTS

Introduction

In this portion I present my conclusions and final thoughts. In the *Conclusions* portion, I present all of the propositions (the SP and the EP) as a summary of conclusions. In the *Final Thoughts* portion, I present, in a less formal and more personal manner, some of final views concerning the SP and the EP.

It might seem redundant to present all of the SP again, given they were already presented in the second chapter of this work and presented throughout. However, there is a difference. First, these propositions are now accompanied by their respective supporting propositions. Second, the SP at the very beginning of this work are mere assertions without any context, i.e., without analysis. Whereas, presenting them in this portion of this work, they are not just mere unexplored propositions without context, but assertions that were properly investigated and elaborated upon in a systematic way; hence, they are now properly understood conclusions in themselves. The EP are propositions that emerged as conclusions by virtue of applying the SP, although, some were elaborated upon as well. I present them together as a summary, given they are now properly contextualized. Consequently, while the conclusions (the SP and the EP) presented here in this chapter are in the form of standalone propositions, they should, nevertheless, be understood within the context of how they came to be within this work; not do so, would misrepresent them.

Conclusions

The SP (this now includes the supporting propositions) have been asserted and investigated and established as true in Chapter 2. I present all seven proposition here as final conclusions:

(P1) Aliens would be more intelligent than us.

(P2) Alien technology would be superior to our technology and necessarily different.
(P2A)Alien technology would necessarily be different from ours due to their world view.
(P3) Aliens would most likely be transgenic beings.
(P4) Aliens would be able to communicate with us clearly.
(P4A) It is incumbent upon the more advanced species to figure out how to communicate with the less advanced species.

(P5) The aliens' own interests would be their primary concern.
(P6) Aliens would owe us no allegiance.
(P6A) Just because the aliens would have learned how to cooperate amongst themselves does not necessarily mean they would cooperate with us.

(P7) Aliens would have good reason to avoid formal public contact with humans.
(P7A) Given how humans treat each other, there is no reason to assume that humans would treat aliens any differently.

Additional conclusions (i.e., the EP) have emerged as a result of the application of the SP within the various properties of friendship in Chapter 3 and therefore are true. I present them here as final conclusions:

(EP1) The property of *communication* would not exist between aliens and humans.

(EP2) The property of *equality* would not exist between aliens and humans.

(EP3) The property of *goodwill* would not exist between aliens and humans.

(EP4) The property of *loyalty* would not exist between aliens and humans.

(EP5) The property of *trust* would not exist between aliens and humans.

(EP6) The property of *honesty* would not exist between aliens and humans.

(EP7) The property of *deception* would necessarily exist between aliens a humans in the extreme.
(EP7A) Just because the aliens could be unfriendly towards us does not necessarily entail our immediate destruction or any destruction of our existence as a species.

These propositions should be understood as a set of principles that can utilized as a tool for the purpose of reconciling the behavior of an alien civilization that would arrive on our planet. Indeed, the SP and the EP can be understood as a synopsis of what one should expect from a Type-III civilization as defined in the Rudnyk scale. It should be noted that out of the seven properties of friendship, only *deception* would exist between aliens and humans. Consequently, it is reasonable to conclude that genuine friendship between aliens and humans would not exist.

Final Thoughts

During the course of writing this philosophical analysis, I was secretly hoping that I would come up with arguments that would have weakened the positions of most of these propositions, especially with respect to the EP. On a personal level, genuine friendship with an alien species is something that is pleasant and inspiring to fantasize about. Admittedly, I still do. It is much more pleasant to think about after all; a guilty pleasure of mine, if you will. However, the analysis of all of the propositions is clear. It is extremely disconcerting that out of the seven key properties of friendship between our two civilizations as explored here—*communication, equality, goodwill, loyalty, trust, honesty, and deception*—that only one would be present, and that is *deception.* To come to the realization through this rigorous process of systematic analysis that, other things being equal, aliens would have no motivation in dealing with us as equals in any fashion, and that deceptive behavior in the extreme from them would be the order of the day, is unsettling. An exception might be if we were to capture some of the beings from a crash, for example. Even then, it is reasonable to assume that they would be cautious with what they would divulge. While all of the propositions are sobering, for me (EP1), (EP7), (EP7A), (P6), (P6A), and (P7) are especially so. Why should we expect anything different from an intelligent and sophisticated civilization? Also, the issue of communication as a non-issue was a surprise to me. Before writing this book, I entertained the thought that an alien civilization would be so different and advanced from us that they might not perhaps know how to communicate with us, and that it would require great effort on their part in figuring out how to communicate with us. However, given what has been explored and concluded in this work, an alien civilization would be able to communicate with us clearly and quite easily. Yet, it is evident that the aliens would have little, if any, motivation in actually formally communicating with us. I have come to the realization that to assume that the arrival of aliens would necessarily entail open arms and open hearts in which they would willingly and openly share their wondrous scientific and technological knowledge in the context of developing a genuine friendship with us is arrogant and naïve. I am not suggesting they would be malevolent. The malevolent or benevolent perspective is the wrong approach in endeavoring to understand the behavior of an alien civilization. It is best to understand the aliens in the context of perceiving them as animals asserting their rightful place within the universe doing what all animals do so that they may live, propagate, and advance; not just to survive but to thrive. We must do the same. And we must strive in being open to ideas and experiences we could not have yet to even imagine for better or worse. However, in the meantime, given the SP and the EP, the bottom line is this:

The arrival of extraterrestrials would not be a day of celebration but of sobriety.

WORKS CITED

"10 Leading Causes of Death by Age Group, United States – 2013." *Centers for Disease Control and Prevention*. <https://www.cdc.gov>. Web.

Aristotle. "Metaphysics: Book I. Chapter 5" *The Complete Works of Aristotle, Vol.2*. Edited by Jonathan Barnes. Princeton University Press, 1984, pp. 1552-1569. Book.

---."Nicomachean Ethics: Book VIII & Book IX." *The Complete Works of Aristotle, Vol.2*. Edited by Jonathan Barnes. Princeton University Press, 1984, pp. 1825-52. Book.

Bates, Mary. "Meet 5 "Zombie" Parasites That Mind-Control Their Hosts." *National Geographic*. <http://news.nationalgeographic.com/news/2014/10/141031-zombies-parasites-animals-science-halloween/>. Published: November 2, 2014.Web.

"By the Numbers: World-Wide Deaths." *National WWII Museum, New Orleans*. <http://www.nationalww2museum.org >. Accessed: 2016. Web.

Cattell, Raymond. "Raymond B. Cattell." *Human Intelligence*. <http://www.intelltheory.com/rcattell.shtml>. Published: 2010; Modified: December 20, 2016.Web.

Clark, Andy and David Chalmers. "The Extended Mind." *Philosophy of Mind: Classical and Contemporary Readings*. Edited by David Chalmers. Oxford University Press, 2002, pp. 643-651. Book.

Colla, Phillip. "Photo Gallery: Electroreceptive Fish." *National Geographic*. <https://www.nationalgeographic.com/environment/oceans/photos/electroreceptive-fish/>. Published January 29, 2010. Web.

Collinson, Diané. *Fifty Major Philosophers: A Reference Guide*. Routeledge, 1987. Book.

Debertolis, Paolo, Fernando Coimbra, and Linda Eneix. "Archaeoacoustic Analysis of the Hal Saflieni Hypogeum in Malta." *Journal of Anthropology and Archaeology*, vol. 3, no. 1, pp. 59-79. *Academia.edu*, https://www.academia.edu/. Web & Journal as PDF.

"Dinocampus coccinellae."*Wikipedia*. <https://en.wikipedia.org/wiki/Dinocampus_coccinellae>. Accessed: August 21, 2017. Web.

Dobkin, David S., Paul R. Ehrlich, and Darryl Wheye. "Brood Parasitism." *Stanford University*. < https://web.stanford.edu/group/stanfordbirds/text/essays/Brood_Parasitism.html>. Published: 1988. Web.

"Flight to Survive: Saving the Whooping Crane." *NFWF* (National Fish and Wildlife Foundation). <http://www.youtube.com>. Published: October 4, 2012. Web.

Fraser, Chris. "Mohism." *Stanford Encyclopedia Philosophy.* <https://plato.stanford.edu/entries/mohism/>. Published: Monday October 21, 2002; Revised: March 23, 2010. Web.

Friedman, Stanton. *Flying Saucers and Science*. New Page, 2008. Book.

Gregg, Justin. Interviewed by Emily Anthes. "Sorry, dolphins aren't uniquely intelligent: The trouble with believing in the special smarts of one charming animal." *The Boston Globe.* <https://www.bostonglobe.com>. Published: February 2, 2014. Web.

"Grow Carnivorous Plants."< http://www.carnivorousplants.org/grow>. *The International Carnivorous Plant Society*. Accessed March 2, 2018.Web.

Guthrie, W.K.C. "Pythagoras and Pythagoreanism." *Encyclopedia of Philosophy, Vol. 7.* Macmillan and Free Press, 1967, pp. 37-39. Encyclopedia.

Huffman, Carl. "Pythagoras." <https://plato.stanford.edu/entries/pythagoras/>. *Stanford Encyclopedia of Philosophy*. Published: February 23, 2005; Revised: May 28, 2014.Web.

---. "Pythagoreanism." <https://plato.stanford.edu/entries/pythagoreanism>. *Stanford Encyclopedia of Philosophy*. Published: March 29, 2006; Revised: October 31, 2014.Web.

Iamblichus. *Life of Pythagoras or Pythagoric Life, accompanied by Fragments of Ethical Writings of Certain Pythagoreans in the Doric Dialect, and a Collection of Pythagoric Sentences.* Translated by Thomas Taylor. J. M. Watkins, 1818. Book as PDF.

Kaku, Michio. "The Physics of Interstellar Travel." <http://mkaku.org/home/articles/the-physics-of-interstellar-travel/>. Accessed 2017. Web.

Kardashev, N.S. "Transmission of Information by Extraterrestrial Civilizations." *Soviet Astronomy-AJ. Harvard University,* vol. 41, no. 2, 1964. pp. 217-21. Retrieved from: *The SAO/NASA Astrophysics Data System.* <http://articles.adsabs.harvard.edu>. Web & Journal as PDF.

Kolak, Daniel and Garret Thomson. *The Longman Standard History of Ancient Philosophy.* Pearson Longman, 2006. Book.

"Korean War." *Wikipedia.* <https://en.wikipedia.org/wiki/Korean_War>. Accessed: June 11, 2017. Web.

Kuhn, Thomas. "The Structure of Scientific Revolutions." Key excerpts thereof. *Introduction to the Philosophy of Science*. Edited and Commentary by Arthur Zucker. Prentice Hall, 1996, pp. 158-167. Book.

Lange, Karen. "Desperate Measure." *National Geographic*. <http://ngm.nationalgeographic.com/2008/09/elephant-management/lange-text>. Published: September 2008. Web.

Loy, Hui-chieh. "Mozi (Mo-tzu, c. 400s-300s B.C.E.)." *Internet Encyclopedia of Philosophy*. <http://www.iep.utm.edu/mozi/>. Published: November 7, 2007. Web.

Ludwig, Theodore M. *The Sacred Paths: Understanding of the Religions of the World, 3rd Edition*. Prentice Hall, 2001. Book.

Matsuzawa, Tetsuro. "Videos." *Chimpanzee AI: Primate Research Institute, Kyoto University*. <http://langint.pri.kyoto-u.ac.jp/ai/>. Accessed 2016. Web.

Mei, Y.P. "Mo Tzu." *Encyclopedia of Philosophy, Vol. 5*. Macmillan and Free Press, 1967, pp. 409-10. Encyclopedia.

More, Max. "H+: True Transhumanism." *Metanexus*. <http://www.metanexus.net essay/h-true-transhumanism >. Published: February 5, 2009. Web.

More, Max. Interviewed by Adam Ford. "Max More – Transhumanism and the Singularity." *Adam Ford*. <http://www.youtube.com>. Published: January 3, 2012. Web.

More, Max. "The Extropian Principles." *HighExistence*. <http://highexistence.com/the-extropian-principles/>. Accessed 2016. Web.

Mozi. "Chapter 2-Mozi"; Introduction and commentary by the authors along with key excerpts from all of the ten doctrines from the *Mozi*." *Readings in Classical Chinese Philosophy, 2nd Edition*. Edited and Commentary by Philip J. Ivanhoe and Bryan W. Van Norden. Hackett, 2001, pp. 59-113. Book.

"Our Work." *Operation Migration*. <http://operationmigration.org/Protocol.html>. Published 2014. Web.

Patuxent Wildlife Research Center. < https://www.pwrc.usgs.gov/>. Accessed: July 30, 2015. Web.

"Philosophy of Language in Classical China."<http://www.philosophy.hku.hk/ch/lang.htm>. *The University of Hong Kong*. Accessed: 2008. Web.

Plato. *Republic*. Translated from the New Standard Greek by C.D.C. Reeve. Hackett, 2004. Book.

"Raising Crane." Season 1, Episode 2. Host & Narrator Mike Rowe. *Somebody's Gotta Do It.* CNN. October 15, 2014. Television.

Reese, William L. *Dictionary of Philosophy and Religion, Expanded Edition*. Entries for "Pythagoras" and "Pythagoreanism." Humanity Books, 1996, pp. 622-4. Dictionary.

Russell, Bertrand. *Wisdom of the West*. Doubleday & Company, Inc., 1959. Book.

"Scientific Goals and History." *The Gorilla Foundation.* <http://www.koko.org/scientific-goals-history />. Accessed: 2017. Web.

"Silk." Season 21, Episode 273, Segment B. *How It's Made*. Discovery Channel. 2013. Television.

Sternberg, Robert J. *Cognitive Psychology, 4th Ed.* Thompson Wadsworth, 2006. Book.

The Buddha. "The Ten Precepts." *Access to Light.* <https://www.accesstoinsight.org/ptf/dhamma/sila/dasasila.html>. 2015. Web.

"The Officer's Professional Reading Guide Top 100 List." *USMA* (United States Military Academy).<http://www.usma.edu/history/siteassets/sitepages/navigation%20bar/oprg%20top%20100%202011.doc>. Accessed: January 24, 2016. Web.

The Upanishads. Introduction and Translation by Juan Mascaro. Penguin Group, 1965. Book.

"Transgenic." *Merriam-Webster*. <http://www.merriam-webster.com>. Accessed: 2016. Web.

Tzu, Lao. *Tao Te Ching*. Translation by Charles Muller. Introduction and Notes by Yi-Ping Ong. Barnes & Noble Classics, 2005. Book.

Tzu, Sun. *The Art of War.* Translation, Preface, and Introduction by Thomas Cleary. Shambhala, 1988. Book.

"Vietnam War Casualties." *Wikipedia*. <https://en.wikipedia.org>. Accessed: 2016. Web.

"Who discovered electricity?" *UCSB Science Line. University of California, Santa Barbara*. <http://scienceline.ucsb.edu/getkey.php?key=4026>. Published: 2015. Web.

Wilcox, Christie. "Biological Warfare: Parasitic Wasp Uses A Virus To Control Its Host." *Discover Magazine*. <http://blogs.discovermagazine.com/science-sushi/2015/02/10/biological-warfare-parasitic-wasp-uses-virus-control-host/>. Published: February 10, 2015. Web.

"World War I Causalities." *Wikipedia*. <https://en.wikipedia.org>. Accessed: 2016. Web.

"World War II Casualties." *Wikipedia*. <https://en.wikipedia.org>. Accessed: 2016. Web.

SPECIAL THANKS

My sincerest gratitude to my loving wife Kathy for her incredible patience & support throughout the whole process of this project. I also would like to thank my mother for her encouragement of getting my book out.

ABOUT THE AUTHOR

Adrian Rudnyk earned an M.A. in Philosophy (Thesis: *"Reconciling Realism with Underdetermination in Musical Ontology"*) from California State University, Los Angeles, and he received a "Special Recognition Award in Graduate Studies". Before that Rudnyk earned a B.A. in Philosophy (Emphasis: Science & Society) from California State Polytechnic University, Pomona, as well as an A.A. in Music and an A.S. in Computing & Information Systems from Citrus College, Glendora. He attended Ukrainian School on Saturdays in L.A. from 1970-1979. Rudnyk had great interest in music and took private violin lessons for about 14 years, played violin professionally, and has taught violin for 21 years. Rudnyk has been a UFO researcher since the mid-1980s, and employs philosophical methodologies as his main strategy in studying the UFO phenomenon. In his leisure time, Rudnyk enjoys: practicing the violin and composing pieces for unaccompanied violin, listening to music, watching T.V., games (i.e., vintage board games, 1980s video games, UNO), Sudoku puzzles, and taking walks with his wife, Kathy.

Made in the USA
Monee, IL
25 March 2026

46848483R00057